SPORTWORKS

More than 50 Fun Games and Activities
that Explore the Science of Sports

SPORTWORKS

from the
Ontario Science Centre

Illustrated by
Pat Cupples

Addison-Wesley Publishing Company, Inc.
Reading, Massachusetts   Menlo Park, California   New York
Don Mills, Ontario   Wokingham, England   Amsterdam   Bonn
Sydney   Singapore   Tokyo   Madrid   San Juan

Library of Congress Cataloging-in-Publication Data

Sportworks: more than fifty fun games and activities
    that explore the science of sports.

    Includes index.
    Summary: Describes a variety of games and activities with
which to explore the science of sports, in such areas as body types,
muscles, and water sports.
    1. Games—Juvenile literature. 2. Sports sciences—Juvenile
literature. [1. Sports sciences. 2. Games] I. Cupples, Pat, ill.
II. Ontario Science Centre.
GV1203.S695   1989   796   88-34317
ISBN 0-201-15296-7

First published as *How Sport Works* in Canada in 1988 by Kids Can
Press, Ltd., Toronto, Ontario.
First published in the U.S.A. in 1989 by Addison-Wesley
Publishing Company, Inc., Reading, Massachusetts.

Book design by Michael Solomon
Cover artwork copyright © 1989 by Arnie Ten
Set by Compeer Typographic Services Limited

ABCDEFGHIJ-MA-89
First printing, January 1989

The Ontario Science Centre in Toronto, Canada, is a vast science
arcade of three connected buildings filled with more than 800
exhibits. If you visit, you can play with them, test yourself against
them, and experiment with them, just as one million other visitors
do each year. The Centre's aim is to let you explore, experience,
and enjoy science.

Although a visit to the Ontario Science Centre is unique, science is
all around us. You don't need a museum or a laboratory to discover
it. You can have the Ontario Science Centre spirit at home with
Sportworks. All you need for the activities in this book are a little
curiosity and some equipment you might already have around your
house or backyard. So have fun!

Other books in this Ontario Science Centre series available from
Addison-Wesley Publishing Company:
    Scienceworks
    Foodworks

Series Editor: Carol Gold
Writers: Carol Gold
                Hugh Westrup

All Ontario Science Centre projects are the product of the
entire staff, whose help in producing this book is gratefully
acknowledged.

CONTENTS

THE SHAPE OF SPORTS

YOU'VE probably never seen a 2 m (6 foot, 7 inch) jockey. And there aren't too many skinny weight-lifters around, either. Tall basketball players, hefty football linebackers, slight, lithe gymnasts—picture the sport and you picture the athlete. Obviously, certain sports offer an advantage for people of a certain size and shape. But do all sports?

Are most swimmers a different shape from most shot-putters? How about rowers and wrestlers? Is there a difference between sprinters and marathoners? Where would you fit in?

Body Type Chart

There are three basic body shapes: endomorphs are round, mesomorphs are muscular and ectomorphs are lean and long. Actually, everyone is a mixture of all three, in differing proportions.

Think of it like mixing paints. You've probably discovered that just a few colours can produce many different shades. You can look at a colour your friend has mixed and guess that it has this much blue and that much red and perhaps just a tad of yellow. If you had the right instruments, you could measure the amount of each colour exactly. It's the same with body shapes.

Look at the athlete in the picture. You could say he has a lot of endomorph and almost as much mesomorph and just a drop of ectomorph.

In fact, that's how researchers do describe body type, as a sort of formula. Of course, researchers use specialized equipment to scientifically measure athletes for height, weight, length, width, circumference and fat. They're then assigned a rating from 1 to 7 for the amount of each component in their body mix. On this scale, 1 is the lowest and 7 is the highest.

The athlete in the picture, for instance, would be a 5-6-2—meaning that judged on the endomorph scale, he's a 5; on the mesomorph scale, he's a 6 and if you look at him from the ectomorph point of view, he's only a 2.

When researchers measured Olympic athletes, they found that, yes indeed!, the average swimmer had a different body "formula" from the average middle-distance runner. It turned out that each sport had its own characteristic body shape. That doesn't mean that all sprinters look exactly alike. But sprinters tend to look more like other sprinters than they do like pole vaulters. It seems that your body shape can give you an advantage in one sport over another.

So what kind of body type are you? Here's how to get a general idea.

You'll need:
several sheets of newspaper
tape
scissors
a marker
a friend

1. Lay out the sheets of newspaper and tape them together. Use enough to make one big sheet large enough for you to lie on.
2. Lie down on the taped sheets and position yourself as in the drawings of the three basic body types.
3. Have your friend trace around you with the marker.
4. Cut your shape out of the newspapers.
5. Rate your shape from 1 to 7 on each of the three types. The more you resemble a type, the higher the score. It's a good idea to ask someone else to rate your cut-out shape, too, for comparison.
6. When you have your three-number body-type recipe, see where you would fall in the Body Type Chart.

If you're still growing, you might want to save your paper shape for a year or two and see how your body changes.

ABS, PECS, TRAPS, LATS AND DELTS

DELTOIDS, pectorals, trapezius and brachialis . . .

Is all of this Greek to you? Well, you're partly right. Most of the muscles in the human body have Greek or Latin names.

Athletes usually get so friendly with their muscles that they refer to them by their abbreviations: delts for deltoids, pecs for pectorals, and so on.

There are more than 600 skeletal muscles in the human body. The major ones are shown here. See if you can find them in your own body.

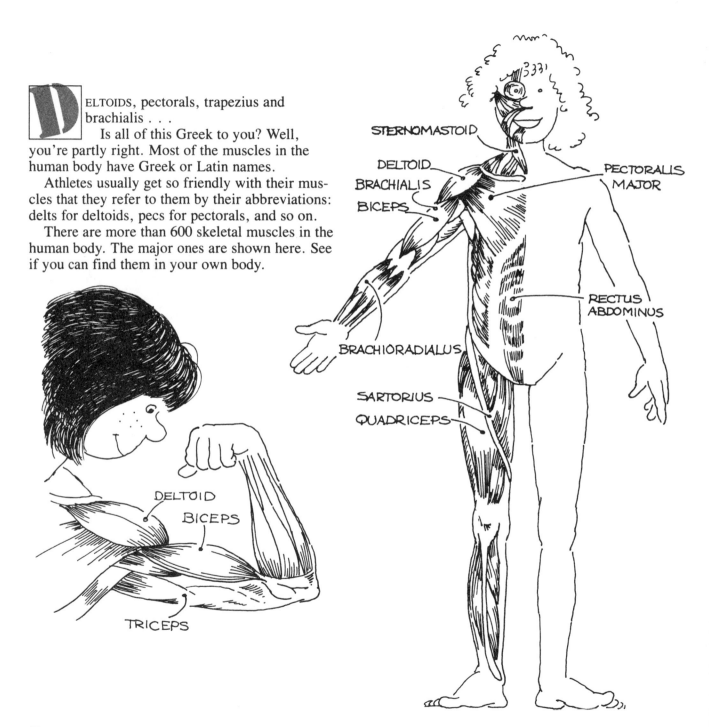

STERNOMASTOID

DELTOID

BRACHIALIS

BICEPS

PECTORALIS MAJOR

BRACHIORADIALUS

RECTUS ABDOMINUS

SARTORIUS

QUADRICEPS

DELTOID

BICEPS

TRICEPS

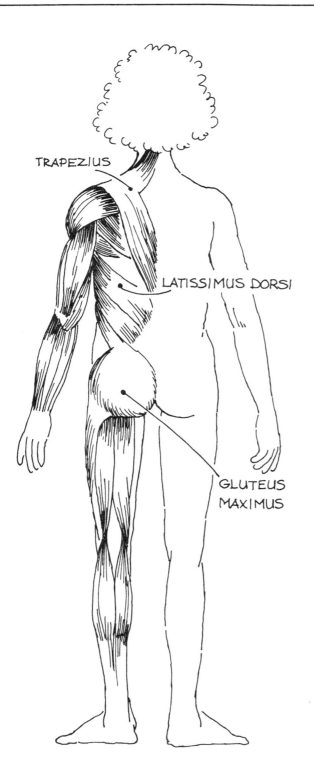

TRAPEZIUS

LATISSIMUS DORSI

GLUTEUS MAXIMUS

Five fascinating facts about muscles

1. About 40 per cent of your body weight is muscle.
2. Your muscles are 75 per cent water! What's the rest? Twenty per cent is protein and the rest is assorted stuff such as salts, minerals and carbohydrates.

LEVATOR LABII SUPERIORIS ALAEQUE NASI

3. One of the smallest muscles has one of the longest names. It's Levator labii superioris alaeque nasi. With a name like that, it should lift mountains, but it doesn't. It's the tiny muscle beside your nose that helps raise your lip into a sneer.
4. For their size, the muscles that operate the wings of bees, flies and mosquitoes are stronger than any human muscles.

5. Ever feel the hair standing up on your arms or neck? It's raised by tiny muscles pulling on each hair follicle.

I LEFT MY MUSCLE IN SAN FRANCISCO

OU and Arnold Schwarzenegger have something in common. You both pump iron.

That's right, the mineral iron is constantly being pumped through your arteries and veins, as it is through Arnold's. And the muscle that does all the pumping is the heart. Just like the biceps in your arms or the quadriceps in your legs, your heart is a muscle.

The heart is the body's strongest muscle. And no wonder. Every day, it flexes about 100 000 times, pumping 12 000 L (3 170 gallons) of blood. In one year, your heart pumps enough blood to fill a supertanker.

Pick up a tennis ball and squeeze it. The force that your hand uses to squeeze the ball is about equal to the force that it takes for your heart to pump iron (and all the other blood components) through your system.

Pumping greater amounts of weight with your arms makes your biceps grow bigger and stronger. The same thing happens to your heart. The more blood it pumps during a vigorous workout, the bigger and stronger it grows. Some athletes have hearts that can pump twice as much blood as an average person's heart.

A well-exercised heart also doesn't need to work as hard throughout the day because it can pump more blood with every beat. There are athletes in such good condition that their regular heartbeat is down around 40 beats per minute. When tennis champion Bjorn Borg was at the height of his career, he reportedly had a resting heart rate of 27 beats a minute. How fast does your heart beat?

Hear a Heartbeat

You'll need:
cardboard tube
a watch that shows seconds
a friend

1. Press one end of the cardboard tube against your friend's chest. Put your ear to the other end and listen to the constant rhythm—"lub, dub; lub, dub; lub, dub."

2. Time your friend's heart rate. Every "lub, dub" equals one beat. Count the number of beats in 10 seconds and then multiply by six. Girls and boys have a heart rate of about 80 beats a minute. Adults have a slightly slower rate—about 70 beats a minute. A baby's heart trips along at 130 times a minute. As you get older, and your metabolism slows down, your heart doesn't need to pump as quickly.

See a Heartbeat

You'll need:
a bathroom scale—not a digital one, but one with an arrow and a spinning dial

1. Do a couple of dozen jumping jacks to get your heart racing.
2. Step on the scale and look closely at the dial. What do you see?

The small vibrations of the needle are in time with your heart.

Your blood is pumping so hard through your body that it makes the scale shake.

Why does the heart speed up during exercise? Your circulating blood carries oxygen to your muscles and other parts of your body. Just as in an engine, your muscles need oxygen to help them burn fuel (fats and carbohydrates). Naturally, when your muscles work harder they require more oxygen, and the heart pumps faster to supply it.

At rest, only about 20 per cent of your blood is distributed to your muscles. During exercise, they receive about 80 per cent.

Calculate a Heartbeat

What's the fastest your heart can beat?

220 minus your age = your maximum heart rate per minute

Any physical activity that pushes your heartbeat up to 70 per cent of its maximum rate, and keeps it there for 15 to 30 minutes, is good for your heart.

Animal	Heartbeats per minute
Shrew	1200
Canary	1000
Mouse	650
Porcupine	280
Golden Hamster	280
Chicken	200
Chihuahua	120
House cat	110
St. Bernard dog	80
Giraffe	60
Tasmanian Devil	54 - 66
Kangaroo	40 - 50
Tiger	40 - 50
Elephant	25
Beluga Whale	15 - 16
Gray Whale	8

BOYS & GIRLS

DID you ever dream of swimming the English Channel? Or one of the Great Lakes? What would give you a good chance of success? Here's a clue: read over the names of these famous ultra-long-distance swimmers.

Marilyn Bell—across Lake Ontario
Irene van der Laan—double crossing of the English Channel
Cindy Nicholas—across Lake Ontario and the English Channel (19 times!)
Shelley Taylor—fastest around Manhattan Island
Julie Ridge—6 times around Manhattan Island
Gertrude Ederle—English Channel
Florence Chadwick—English Channel

What do these people have in common?

It's not an accident that they're all female. Many long-distance swimmers are. Why does being female give them an advantage?

To find out, pinch your mom on the outside of her upper arm—gently! Now, pinch your dad in the same place. What you'll probably find is that mom has a thicker layer of fat under her skin.

Men have more muscle than women—about 20 per cent more—but it takes more than strength to cross the English Channel. Women have a natural advantage in long-distance swimming because they have more fatty tissue, which is stored in their breasts, under their skin, on their hips, buttocks and elsewhere.

Why does extra fat help you swim?

Greater body fat provides women with an additional source of fuel and serves as an extra layer of insulation, which comes in handy when you're striking through the frigid waters of the English Channel.

Also, fat acts like a built-in life preserver, making it easier for women to stay afloat.

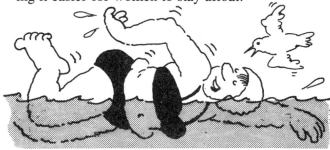

Finally, body fat gives a woman her rounded shape. Having a rounded shape is beneficial for swimming because it makes it easier to move through the water—a woman's bones don't stick out and slow her down as much as a man's. As a result, women use less energy when they swim.

Nature equipped women for long-distance or endurance sports. Men are more naturally built for strength and speed.

However, that difference doesn't exist in kids under 12. Boys and girls have equal amounts of fat and muscle and are about the same size. It's only when puberty arrives and various hormones start flowing that the differences begin to take shape.

Flexibility Test

Now that you've pinched your parents (or any other handy adults), they're probably paying attention. Ask them to join you in this flexibility test.

1. Sit down on the floor with your feet and legs sticking straight out in front of you.
2. Bend forward and touch your toes.
3. Now see how far past your toes you can reach. Measure the distance.

Compare your results of this test with those of the adults. How do you all rank in flexibility?

If you're under 12, you're probably more flexible than any of the adults. That's why the top female gymnasts are in their early teens or younger—they need that extra flexibility to perform the "pipe cleaner" bends required to win in competition. But don't gloat about your flexibility—once you reach puberty, you'll start to lose flexibility, too.

Boys will lose more flexibility than girls. Females have more of a hormone called relaxin, which softens and stretches their ligaments—pieces of tissue like strong elastic bands that connect bones to each other. Because women have more relaxin, their joints are more flexible and can bend farther than men's. With greater flexibility, women can take longer strides in running and better kick and arm movements in swimming.

THE HUMAN PRETZEL

I F you've ever been to a circus or a carnival, you've no doubt gasped at the The Human Pretzel or The Snake Woman. These human wonders are so supple they can tie themselves in knots, or sit on their own head, or rotate their head practically back-to-front.

They are contortionists, sometimes referred to as "people without bones." But the key to their extraordinary elasticity really lies in the soft tissue that covers and links their bones. Called connective tissue, this includes the muscles, ligaments and tendons. From a very young age, contortionists practise hard at stretching their connective tissue until they can extend it beyond all normal limits.

The soft, connective parts of your body are like a series of interconnecting rubber bands. Your body has thousands of "rubber bands" in the form of muscles, tendons and ligaments. Every move you make involves many of these elastic bands

joining together in a team effort, some of them relaxing, others being pulled tight.

Here's how you can feel this going on. Bend your upper arm as though you were "making a muscle." Use your other hand to feel what happens when you do this. The biceps muscle in your upper arm actually tightens up, while the triceps muscle at the back of the arm relaxes. When straightening your arm, the biceps loosen while the triceps contract.

Athletes, particularly gymnasts and figure skaters, put extraordinary demands on their connective tissues. It takes super flexibility to execute a triple axel or perform a back walkover on a balance beam.

Watch the sidelines and you'll see athletes doing 10-15 minutes of simple stretching exercises before a competition. They're limbering up, so they'll have a full range of motion in each joint and won't risk pulling a muscle or a tendon.

The rubber bands in your body won't keep their stretchiness unless you exercise them. Through inactivity, they can grow too tight or too loose, and the results can be stiffness and clumsiness, aches and pains.

How good is your flexibility?

These tests should give you an answer:

Flex test No. 1

1. Stand and cross your legs with one leg in front of the other.
2. Bend slowly at the waist and try to touch the floor in front of your toes. *Don't bounce!*
3. Hold yourself in this position for about five seconds.

If you can do this, you have good flexibility in the muscles at the back of your thighs.

Flex test No. 2

1. Take off your shoes and stand on your heels with your toes lifted off the floor.
2. Walk straight ahead 10 steps.

If you can do this without losing your balance, you have good flexibility in your Achilles' tendons.

Flex test No. 3

1. Sit on a table top with your legs hanging over the edge and the backs of your knees touching the edge of the table. Separate your knees by several centimetres (a few inches).
2. Tuck in your chin and slowly bend forward. Try to lower your head all the way between your knees.

If you can do this, you have good flexibility in your lower back.

CENTRED ON GRAVITY

MEN are better at sports!'' ''Women are just as good!''

That's an argument that all sports fans get into sooner or later. Who wins? Everybody. Men and women are equally good at sport, but each excels in different sports.

That's because men and women are built differently. A man and woman may be of equal height and weight, but the man will have more muscle, while the woman will have more body fat. That means men are more suited for sports that emphasize strength while women do better at sports that require endurance.

It's not just the amount of muscle that matters, it's where the muscle is. Size being equal, men tend to have stronger arms; women have stronger legs. So men can hit or throw baseballs farther but women can make better cyclists.

Another difference is size. Men are, on average, larger than women. In many sports, such as football, large size is an advantage, while in others—gymnastics, for instance, or being a jockey—small size is better.

Another difference between the sexes is in their centre of gravity. You can find out what this difference does.

WHERE'S THE CENTRE OF GRAVITY?

V!

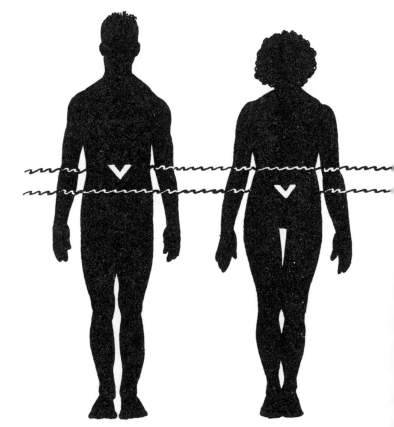

16

Centre of Gravity Test

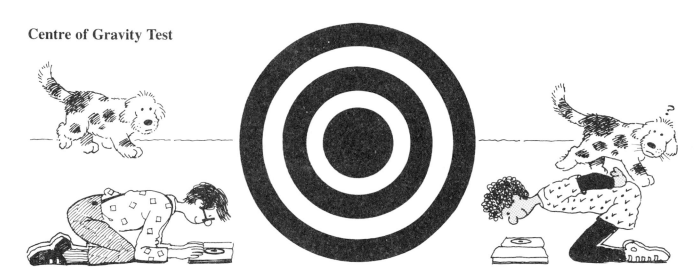

Scientifically, centre of gravity is the point in any object around which its weight is equally distributed. For example: Take a pen and place it across your finger so that it balances there perfectly. The point at which the pen stays without tipping is its centre of gravity.

The same goes for bodies (although they're harder to balance on your finger).

In this test the object is to touch the target on this page with your nose without losing your balance.

1. Put the book on the floor open to this page.
2. Kneel on the floor with your elbows against your knees and your fingertips touching the edge of the target.
3. Straighten up and put your hands behind your back.
4. Bend over and try touching the centre of the target with your nose.

Who's best at this test—men or women?

You can do this test only if your centre of gravity remains in the area directly above your knees and feet. Men tend to have broader shoulders and more muscular chests and arms, while women have broader hips and more weight in the lower part of their bodies. This gives men a higher centre of gravity than women. When a man stretches forward, his high centre of gravity shifts outside the zone above his knees and feet, and he falls over. But because a woman's centre of gravity is lower, it remains within the balance zone, allowing her to reach the target without tipping over.

The closer your centre of gravity is to the ground, the more stable you are. Having a lower centre of gravity is an advantage in sprinting, because you're less likely to be blown over by the wind or to fall over when you stumble.

On the other hand, having a high centre of gravity can be advantageous in a sport such as high jumping, where the trick is to get your centre of gravity over the bar. Naturally, if you've got a high centre of gravity, you can leap over higher bars.

MUSCLES AND CRANBERRY SAUCE

Has anyone ever called you a turkey? Well, in one respect, they were right! Just like a turkey, you have white meat and dark meat. And in sports, that's cause for thanksgiving. If you were to extract some muscle tissue from your body and look at it under a microscope, you'd see a bunch of long, stringy fibres running side by side, like wires in a telephone cable. Some of the fibres would be richly supplied with blood; these make up the dark meat. The other fibres, which don't receive as much blood, would be the white meat.

Since you're not eating your muscles, but using them, you might not want to call them dark or white meat. Instead, you could use the names given them by scientists who study bodily changes during exercise. They call the white meat *fast-twitch fibres* and the dark meat *slow-twitch fibres*.

Fast-twitch fibres contract very quickly and yield a short burst of energy. You use your fast-twitch fibres whenever you make a sudden show of speed or strength. Sports such as shot-putting or the 100 m dash are fast-twitch events.

Slow-twitch fibres, on the other hand, are used in sustained activities that don't require a flat-out display of effort, ones that last for more than about two minutes. Endurance sports, such as the marathon or cross-country skiing, are slow-twitch events. Most sports use a combination of these muscle fibres.

Everyone has the same overall number of fibres in each muscle, though the proportion of slow-twitch to fast-twitch fibres varies from person to person. You may have 80 per cent slow-twitch fibres and 20 per cent fast-twitch fibres, while your best friend has a 40-60 ratio. It's something you're born with and there's little you can do to change it.

Being born with more fast-twitch fibres endows you with a natural advantage in sprinting, while having more slow-twitch fibres gives you a head start in marathon racing. Alas, there's no way to count your fibres without surgically removing some of your muscle tissue.

Which Twitch Is Which?

You can't see your fast-twitch muscle fibres but here's how to feel their presence.

1. Stand with your back pressing against a wall and your feet planted about a step away from the wall.
2. Slide down the wall until you're almost in a sitting position. Hold yourself there for as long as you can. How do your legs feel?

That soreness in your thighs as you hold your position against the wall comes from a buildup of lactic acid, which is produced by fast-twitch muscle fibres when they're working. Lactic acid "gums up" the muscles, and must be constantly flushed out of your system. Sitting against the wall makes your thigh muscles so tense that the lactic acid cannot easily be removed and you feel pain. During heavy exercise, lactic acid can build up in your muscles in large quantities, which explains why you sometimes feel achy immediately after exercise. (The immediate achiness caused by lactic acid is different from the soreness that comes the day after exercise. The "next day" pain comes from microscopic tears in the muscle tissue caused by exercise.)

Warm Down

You've just finished three sets of tennis or several laps around the track and your immediate inclination is to sit down and rest.

Don't.

Follow exercise with a "warm down" period. Go for a five-minute walk, or do some stretching. Winding down gradually after exercise helps your circulation flush the excess lactic acid out of your muscles and reduces post-exercise soreness.

Fast-Twitch Food

It's a pain when lactic acid accumulates in your muscles, but you can eat it without any ill effects. The sour taste in plain yoghurt—that's lactic acid.

KEEP YOUR COOL

EY, you! You're just sitting there reading this book, taking it easy. No sweat, huh? Wrong, damp one. Take a close look.

You'll need:
a strong magnifying glass
a desk lamp

1. Make a fist with one of your hands for 10-15 seconds.
2. Open your hand and hold it, palm side up, under the light. Look carefully at the tips of your fingers through the magnifying glass.

Those tiny pieces of glitter on your fingertips are actually droplets of sweat catching the light.

Now that you've seen that even sitting and reading can make you sweat, how much sweat do you think you produce in a day? Well, if you relaxed in an air-conditioned movie theatre all day, you'd lose about a cup of sweat. Spend a hot day playing sports or just running around and you could produce almost a bucketful. Adult marathon runners can produce nearly three times that much in a single race. If they don't replace all that lost water throughout the competition, the results can be serious, especially if it's a hot day. You can actually lose so much water that your body stops perspiring, which can lead to heat exhaustion or deadly heatstroke.

If you just have to put all this water back, why get rid of it in the first place? Sweating is a water-cooling system. The engines of the human body, the muscles, produce a great deal of heat, which the body must eliminate to keep its temperature from rising to dangerous levels. During heavy exercise, muscles transfer their excess heat to the blood, which then rushes to the surface of the body. That's why some people get so red-looking when they exercise.

Meanwhile, the brain has instructed the pores of the skin to squeeze out droplets of water, which form the layer of sweat on the skin. When the hot blood reaches the surface of the body, its heat is immediately transferred to this sweaty layer. The heat is used up by evaporating the sweat. The more heat your body produces, the more sweat it needs to carry the heat away.

It Ain't the Heat...

You've probably heard a million times the old saying, "It ain't the heat, it's the humidity." Make that a million and one, though it's still true. A muggy day feels worse because the human water-cooling system can't operate efficiently.

When it's dry outside, the air can easily soak up all of the sweat the body produces. But when the air is thick with moisture like a sodden sponge, it hasn't the capacity to accept much more water. And so the perspiration just sits on your body, soaking your clothes and streaming down your brow. Unable to cool off properly, you feel tired and miserable.

To protect themselves from heat exhaustion, professional athletes often shorten their training time on humid days, or they play with lighter equipment. (Baseball players, for example, play with lightweight bats.) And on any hot day, humid or not, you should wear light, loose-fitting clothes that enable you to sweat as freely as possible. Wearing too much clothing can be like stepping into a muggy day—it stops your body heat from escaping into the air.

TRY THIS

Find out the difference clothes can make.

You'll need:
two pots of the same type
a stove with two burners
permission to use the stove

1. Put about 5 cm (2 inches) of water in each pot.
2. Put the pots on the stove and turn the burners to low heat. Cover one pot and wait for about 10 minutes.
3. Turn off the heat and wait about 10 seconds.
4. Hold your hand about 18 cm (1 foot) above each pot and compare the heat that they give off.

Imagine that one pot is you with your clothes off on a hot day. The other pot is you with your clothes on. Which one is which? Why?

THE 20-SECOND WORKOUT

YOU'VE probably seen an aerobic workout. You know. . .designer gym togs, sweat bands and non-stop jumping jacks, bun lifts, froggy-ups and ham sandwiches. But have you ever seen an *anaerobic* workout? What's the difference? Try this test to find out.

You'll need:
a flight of stairs
a stopwatch or a watch with a second hand

1. Run up and down the stairs for 20 seconds. While you're running, notice how you're breathing.
2. After a short rest, run around the block at a steady pace. Notice how you're breathing this time. Was there any difference between the way you breathed in the two exercises?

"Aerobics" is a Greek word that roughly means "with air." Aerobic exercises are so-called because they require oxygen to produce the energy for rapid movement of the muscles. When you ran around the block, did you notice how heavily you were breathing? That's because your body required big gulps of oxygen. You were doing an aerobic exercise.

If aerobics means "with air," then "anaerobics" must mean—you guessed it—"without air." An anaerobic exercise is one that doesn't require oxygen for the production of energy. When you ran up and down the stairs, you didn't breathe as deeply as you did while circling the block. You were doing an anaerobic exercise.

Your body is able to produce energy without oxygen only during brief periods of intense activity. Anaerobic exercises never last for more than about two minutes. Once you pass the two-minute mark, your aerobic system starts to take over.

If you think about it, the anaerobic-to-aerobic system is a very clever mechanism. When the starting pistol fires and you bolt from the starting blocks, or when you meet a big grizzly bear in the bush and must run for your life, your lungs just can't breathe fast enough to meet your muscles' sudden high demand for oxygen. So you rely on a source of energy—anaerobic energy—that doesn't require oxygen, until your breathing catches up to the movement of your limbs.

Which sports are anaerobic? The shot put, the hammer throw and the 100-metre dash, to name a few. Most sports are a combination of anaerobic and aerobic. In basketball, for example, a player running back and forth across the court operates aerobically, but with occasional bursts of anaerobic energy for every jump shot or run at the basket.

Taking out stitches

Have you ever laughed so hard your sides ached? The pain you felt was the same as the kind of pain that you sometimes get in your side when you exercise too hard. Both pains are called stitches.

A stitch is a cramp in your diaphragm, the muscle that controls your breathing. During exercise (or laughter), the human diaphragm functions like any other hard-working muscle; sometimes it can tighten up and fail to relax, like an overworked leg with a charley horse.

The best treatment for a cramp is to stretch the muscle to its normal relaxed position and then massage it. Once the cramp is gone, you can go back to exercising—or laughing— though if the pain returns, it's probably best to stop for the day.

RY THIS

You'll need:
two pieces of paper
a pen
a stopwatch or a watch with a second hand

1. Make a list of 20 three-letter words such as toy, ark, box, she, etc.
2. Make a second list using the same words but with the letters all mixed up so that toy becomes yot, ark becomes kra, box becomes xob and she becomes seh, and so on.
3. Ask a friend to read out loud the first list and time him.
4. Ask him to read out loud the second list and time him again. Which time was faster?

Why?
You respond more quickly to familiar patterns.

What's the point?

Well, it could have a lot to do with being an expert hockey player like Wayne Gretzky. Or a chess grand master. Or a great shortstop. They all learn to recognize the patterns in their games.

To become an expert hockey player, you have to do two things. Number one, you have to sharpen your physical skills—skating, handling the puck, checking. Number two, you have to understand that learning to play hockey is just like learning how to read.

Remember the day you picked up your first book? You opened it, and there, staring back at you, was a jumble of squiggles that made no sense. But with practice, you soon discovered that the squiggles formed letters and the letters gathered together to form words. The nonsense turned into sense.

The same thing happens when you watch your

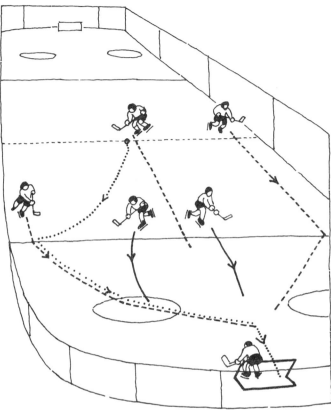

first hockey game. All you can see is a bunch of players scattered across the ice, like jumbled-up letters on a page. But the more you watch, the more clearly you pick out how those players are working together as a team. From one set of positions, the players move to another, to another, and another, just like letters rearranging themselves to form different words.

A hockey team can arrange itself into hundreds of patterns during the course of one game. Just as it takes a long time to build up a good vocabulary of words, it also takes a long time to recognize all the patterns in hockey. But once you've learned the patterns, you always know what to do in every situation—when to pass the puck, when to check an opponent, and so on. Even better, you start to anticipate what will happen next. They say that Gretzky knows where the puck will be three seconds before it gets there.

Scientists estimate that it takes 10 000 hours of practice to recognize all the patterns that players can fall into on the ice. Practice really does make perfect. Expert hockey players are made, not born.

Soccer, basketball, football, chess—each of these is like hockey; the players must learn to recognize the patterns of the game. But there are other sports where patterns are less important.

In volleyball, for example, you will acquire the ability to scan and find the position of the ball. Expert volleyball players don't watch the other players so much as they focus on the speeding ball, which can be flying through the air at speeds of up to 145 km/h (90 mph). Once again, this ability to scan takes thousands of hours of practice.

Whether the game is hockey or volleyball or soccer, it all goes to show you that every sport asks you to use your brain as well as your body.

CROUCH

NCE upon a time, in the late 19th century, in the small town of Springfield, the most important day of the year was Race Day. That was the day all the older boys and young men of the area competed in a race down Main Street. The race was followed by picnics and square dancing. The winner of the race was the star of the day and received the coveted trophy, the Springfield Buffalo.

Jock was ready. He'd been practising all year, and was the fastest fellow in town. He'd won the race the past two years and was looking forward to a third victory. "You fellas can forget ever having that Buffalo sit on your mantcl," he'd boast. "When I win it for the third time, it'll be mine to keep."

Race day dawned sunny and cool. "Just perfect," thought Jock, as he sauntered jauntily toward the starting area, carrying the Springfield Buffalo under his arm with the air of an owner who is bringing a prize possession out for others to admire.

At the starting line, the competitors were warming up. Jock looked around confidently. There was only one new face, a young man named Neil, whose family had just moved to the area. He was quiet and a little shy, perhaps because the others made fun of him for being different; Neil came

from the city and his ways of doing things were sometimes strange. Jock wasn't worried. What could a city boy know about running?

The race was about to begin. The mayor held the trophy aloft and called the competitors to line up at the start. "Here it is, boys! The honour and glory of Springfield goes to the first one across the finish line."

The young men jostled for position and Jock found himself standing next to Neil. Jock prepared himself, standing straight and tall, clenching his fists with tension.

"Get ready!" cried the mayor. And suddenly, to Jock's amazement, Neil dropped into a crouch.

"Hey!" Jock said, pulling at Neil's arm, "Get up! I'm all ready to run!"

"Got a loose shoelace?" the mayor asked Neil.

"No, sir," said Neil politely, freeing his arm from Jock's grasp. "I like to start a race this way."

After a moment's shocked silence, all the competitors and those on the sidelines who had heard started to laugh. Quickly, the word spread: "The new boy thinks you should start a race by bending over!"

Despite the laughter, Neil was determined to do it his way. "Let's get started," said Jock. "If he wants to stay bent over, let him. When I get going, he'll really eat my dust!"

"Get ready," the mayor called out again.

"Get set.

"GO!"

Jock took off, running as fast as he could. He heard the shouts of the crowd and felt a swell of pride. Suddenly he realized they weren't cheering him. Someone was ahead of him. He ran harder, his legs aching. His lungs felt as though they were bursting; he was running too hard to breathe.

The finish line, just ahead! As he stuck his chest out to break the string, he saw it fall, broken by someone who had beaten him. Who? Gasping, he looked around him. Neil was shyly smiling as he accepted the congratulations of the crowd. . .and the Springfield Buffalo.

"How did you do that?" Jock demanded. "You must have started running before the mayor yelled 'Go!' How else could you beat me when you were all bent over at the start?"

"That's exactly how I did beat you," explained Neil quietly. "I've seen the top runners practising in the city and they all start that way now. They say it gets you off faster."

"Well, I'm never going to look like a fool and squat down to start a race."

"You will if you want this back," smiled Neil, cradling the Springfield Buffalo as the crowd lifted him to their shoulders and headed for the picnic.

From then on, all the runners in the big race crouched down at the start. The technique became known in that area as the Springfield Kneel.

The Moral of the Story

There are reasons why Neil's crouching start produces more speed. Each muscle has one position in its range of movement in which it has the most power. You've probably noticed this if you've arm-wrestled. The sprinter's crouch, or "set" position, brings the large muscles of the thighs into their maximum power range for the quickest possible push-off from the starting blocks.

In a race, you want to use your energy to move toward the finish line. When you crouch, most of the pushing power of your legs is directed straight back. That way, you move forward with the least amount of wasted energy.

In a short race, a fast start is very important. A runner who can get to top speed faster than anyone else has an advantage, because there isn't much time for the rest of the racers to catch up.

SPORT MYSTERY 1:
WHY DO CURVE BALLS CURVE?

 AVE you ever seen a ball break in mid-air? It happens all the time, or haven't you noticed?

"Break" is the word that baseball pitchers use to describe those throws that curve in flight—they *break* away from their original path. From a batter's point of view, a curve ball will seem to come straight in and then drop suddenly, as if it had rolled off the edge of a table.

To make a ball break, you have to spin it when you throw or hit it. Why does a spin put a curve on a ball? A ball moving through the air is surrounded by a thin layer of air which "sticks" to the ball as it flows around it. (The stitches on a baseball and the fuzz on a tennis ball help to grab hold of this layer of air.) On a ball without a spin, this layer of air moves around the ball and then separates from it at the back to form a wake of turbulent air.

On a ball with a top spin, however, the turning motion of the ball pulls the air on the bottom around to the back and then gives it an extra shove, throwing the wake upwards. The effect of this is to push the ball towards the ground.

Similarly, the wake on a ball with a back spin will be thrown downwards, keeping the ball from dropping as quickly (a back-spin curve is what creates a fastball). The wake on a ball with a side spin will be thrown to the side, pushing it in the opposite direction.

Being able to put a curve on a baseball (or a tennis ball) takes a strong arm, plenty of practice, proper coaching and a certain level of physical maturity. Many baseball coaches advise kids not to try putting a curve on a baseball until they're in their mid-teens.

But you don't need all of that to put a spin on a beach ball.

TRY THIS

You'll need:
a beach ball

1. Hold the beach ball on your outstretched hand.
2. Bring your other hand up fast, the way you would if you were hitting a volleyball, but instead of hitting the ball right on, let the flat of your hand hit it on the side as you pass it. It's almost as if you were grazing the side of the ball, but hard.
3. Watch it curve!

Foam balls are also good for trying out a few curves. Grip the ball with your thumb and first two fingers. Throw, but just before you let go of the ball, twist your hand slightly to one side to start it spinning. Start out with side spins which are easiest to throw, and then move on to a top-spinning ball. Try throwing it with a back spin against the floor. Which way does it bounce?

ANIMAL ATHLETES

IMAGINE trying to arm-wrestle an octopus! You'd need three friends to help you. Or how about boxing with a kangaroo? You'd want to make sure you were just shadow-boxing because kangaroos fight with their powerful hind legs and with one kick, the kangaroo would have the boxing match in the bag. . .er, pouch. Then again, think about a swimming meet with a dolphin. If it won, you'd know it had beaten you on porpoise.

These matches may sound a bit fishy, but what if humans really did compete in sporting events with animals? Who would be the winners? Could you swim faster than a shrimp, or outrun a rhino? How do people athletes compare with animal athletes?

Weight Lifting

One day at the Bronx Zoo in New York City, a 45 kg (100-pound) chimpanzee lifted a weight of 270 kg (about 600 pounds)—six times its own weight. (A champion Olympic weight-lifter can lift only four times his own weight.) But the gold medal for weight-lifting in the animal kingdom doesn't go to the chimp. How about the ant? Laboratory measurements show that an ant can pick up and carry back to its nest a rock 50 times its own weight. That would be like you picking up a 2.5-tonne elephant and toting it up a steep hill.

But even the mighty ant doesn't win the weight-lifting championship. An ant is a weakling next to the Goliath beetle, which can lift at least 850 times its own weight. If you had that much might, you could carry four school buses on your back. Goliath beetles, which live in tropical climates, are so strong that children hitch them up to toy wagons and hold beetle harness races.

Long Jump

When it comes to the long jump, we humans don't do all that badly alongside the other large mammals. Humans can jump nearly 9 m (approximately 30 feet), and only a few mammals can beat that—horses can jump 11 m (37 feet), mountain lions 12 m (39 feet), and kangaroos nearly 13 m (42 feet).

But it's the small animals that jump farthest in relation to their body size. While an Olympic competitor can jump about 5 times the length of his or her body, a jack-rabbit running to save its life can jump 11 times the length of its body. The jerboa, a small rodent found in Asia and Africa, hops on its hind legs like a tiny kangaroo and can jump 20 times the length of its body. The tiny jumping mouse, which can fit into your hand, can leap more than 30 times the length of its body.

But by far the best long jumper is the flea. With legs that measure only about 0.13 cm (1/20 of an inch) long, a flea can jump 200 times the length of its body. To beat that accomplishment, a human long jumper would have to leap five city blocks.

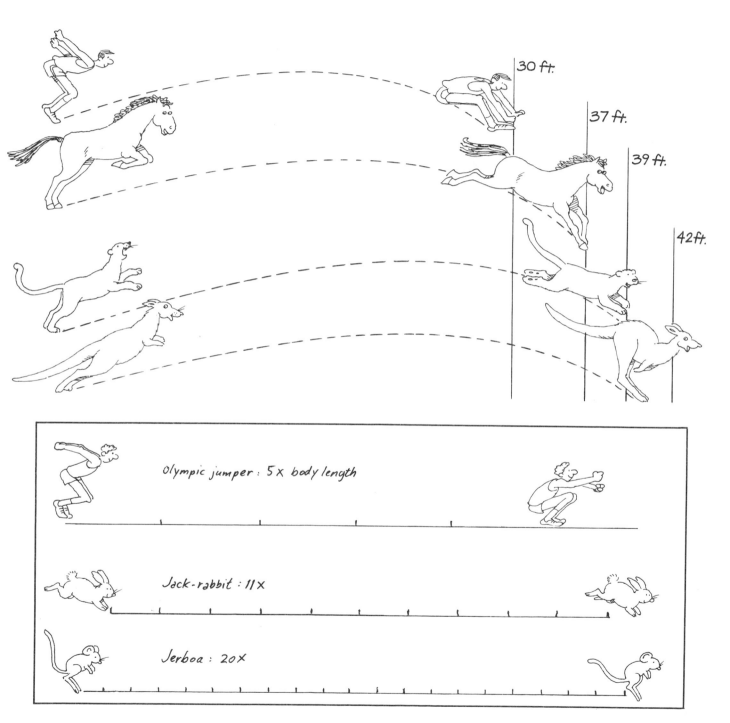

30 ft.

37 ft.

39 ft.

42 ft.

Olympic jumper : 5 x body length

Jack-rabbit : 11 x

Jerboa : 20 x

Swimming

When it comes to swimming, people are like fish out of water. The fastest Olympic swimmer trolls along at a mere 8 km/h (5 mph)—not much faster than the fastest shrimp. (It zips along at just over 3 km/h or 2 mph.) Walruses can get up to about 24 km/h (15 mph) in water, while a leatherback turtle will hit speeds of 35 km/h (22 mph).

The best swimmer among the birds is the penguin (35 km/h or 22 mph) and among the mammals, the dolphin (46 km/h or 29 mph). But the fastest of them all is the sailfish, which lives in tropical and subtropical waters and is most abundant around south Florida and the Galapagos Islands.

The sailfish is instantly recognizable because of its large dorsal fin, which looks like a huge ruffled potato chip. The sailfish can raise and lower this fin like a real sail. The fin stays down when the fish is putting on a burst of speed, then lifts to keep the fish steady. Sailfish have been clocked at a top speed of 110 km/h (68 mph), almost twice the speed of the world's fastest nuclear submarines!

Running

The only way a human could win this event would be in a speeding car. And breaking the speed limit is exactly what you'd have to do to overtake the cheetah, which has been clocked at a top speed of 114 km/h (71 mph). Olympic runners can make it only up to 44 km/h (27 mph).

Few cars could beat a cheetah in a test of acceleration. The sleek, spotted African cat does 0-72 km/h (0-45 mph) in a mere two seconds.

How does a cheetah move so fast? The answer lies in the cheetah's supple spine, which bends like a spring as the animal runs. This enables the cheetah to bring its hind legs well forward of its front legs on each leap. At the beginning of the next leap, the spine straightens out, providing extra thrust to its strong hind legs.

Animal Maximum speed:	Mph	Km/h
Snail	.03	.04
Giant Tortoise	.23	.37
Human	27	43.5
Black Rhinoceros	28	45
Cat	30	48
Grizzly Bear	30	48
Giraffe	37	59.5
Greyhound Dog	42	67.5
Ostrich	43.5	70
Jack-rabbit	45	72.4
Thomson's Gazelle (Cheetah's dinner)	50	80.4
Cheetah	71	114.3

THIS SIDE UP

AVE you ever noticed how basketball players crouch with their feet spread wide apart as if they were sitting in a chair? Is it because they're shy about their height? No, it's to maintain their balance.

Because they must react quickly to the movements of other players, basketball players are constantly at risk of losing their balance and falling over. Part of learning basketball, or most other sports, is sharpening the body's natural sense of balance so that you stay upright at all times. So the basketball player stands with feet far apart, while the skier leans forward and crouches down, and the boxer moves with one foot in front of the other.

You probably don't think much about your sense of balance from day to day, though life is really one long balancing act. Every movement throws you at least a little off centre, putting you in danger of toppling over.

TRY THIS

You'll need:
a full-length mirror
a ball of string
tape

1. Cut a piece of string the length of the mirror.
2. Tape one end to the centre of the mirror at the top and let it hang straight down. Tape the lower end of the string in place at the bottom of the mirror.
3. Stand in front of the mirror with your feet together and one eye closed and line up your nose with the string.
4. Lift your left leg just a little. Where does your nose go?
5. With your leg still hanging in the air, lift your right arm until it sticks straight out beside you. Where does your nose go now?

When you lifted your leg, your body was thrown off balance. That's why your nose (and the rest of you) shifted to the right—to keep you from falling over. Raising your right arm provided a counterbalance for your raised leg, and so your body moved back toward the centre.

Your arms and legs work to counterbalance each other just like two people on a teeter-totter. But because your legs are much heavier than your arms, it's like a fat kid and skinny kid sitting on the teeter-totter. Just as the skinny kid must sit farther out than the fat kid to keep the teeter-totter in balance, your arms must stick out farther than your legs to keep your body from tipping too far to one side.

How does your body keep adjusting its balance when you're always moving your legs and arms or shifting positions? By paying attention to a constant flood of messages from your eyes, joints and muscles. Like planes around a control tower, the eyes, joints and muscles are constantly letting your brain know where your limbs are located in relation to each other, and where you stand with respect to the ground.

But to make it all work, you need rocks in your head. And lucky for you, you have 'em. The rocks are actually tiny crystals and they're in two little hollow sacks, one near each ear-drum. The sacks are lined with microscopic hairs. Whenever your body accelerates forward, or your head tilts to one

side, gravity causes the rocks to move, pressing against different hairs in the lining. Fast as lightning, the hairs relay a message to your brain, telling it about your change in position.

Are you well-balanced?
You'll need:
a watch with a second hand or a stopwatch
a friend

The Blind Stork Test
1. Put on running shoes and stand on a hard surface (not a rug). Get your friend ready with the watch.
2. Stand on your dominant leg (the one you kick with) and press your other foot against the knee of the leg you're standing on. Put your hands on your hips. Close your eyes.
3. As soon as your eyes are closed, your friend should start timing you with the watch. How long can you stand without shifting your foot or taking your hands off your hips or your foot off your knee?

This is a test of "static" balance. If you did well, you might make a good high diver or gymnast.

Hop till you Drop
1. Stand in stocking feet on a slick floor (e.g., tile or linoleum). Take the same position as for the first test but keep your eyes open this time.
2. Have your friend start the watch and tell you when five seconds has gone by. Then make a half turn by swivelling on the ball of your foot.
3. Keep turning every five seconds until you take your hands off your hips or your foot off your knee.

This is a test of "dynamic" balance. It shows how fine-tuned the muscle receptors in your legs are. If you did well, you might make a good surfer or downhill skier.

MEASURE YOURSELF

ID you ever go through all the jeans in a store looking for a pair that fit just right? Some of them need only a washing or two to shape themselves to you. Others are never going to fit your particular body, no matter what you do. You're just not the shape the jeans were designed for.

The same is true with sports. Sometimes, your natural shape and abilities aren't what a certain sport was designed for; other times, it's as though the sport were sewn just to fit you.

In between, there are all sorts of clothes—er, sports—that may not be an exact fit, but that look pretty good with a little altering or a few accessories.

So how do you know if a sport fits you? You could try it on—work at it for a while, see if you're good at it. Or you could, in a way, hold it up to yourself and see—roughly—if it matches your natural shape and abilities.

How can you "hold it up to yourself"? With a bit of measuring!

Try Your Lungs on for Size
You'll need:
a big plastic bag (such as a kitchen garbage bag)
a marker that will write on the bag
a funnel
a container (such as a pitcher) marked off in litres (quarts)

1. Bunch together the opening of the bag to make a mouthpiece, the way you would with a paper bag you were planning to blow up and burst. Make the mouth opening wide so you can breathe into the bag with your mouth open.
2. Squeeze the bag to get the air out.
3. Hold the bag away from your mouth and take two normal slow breaths.

4. On the next breath, breathe in as much air as you can, then bring the plastic bag to your mouth.
5. Pinch your nose and breathe out hard in one breath into the bag. Keep your mouth open— don't purse your lips as though you were blowing. Continue pushing the air out until you feel as though every last drop of air is squeezed from your lungs. (Hint: It helps to bend forward as you breathe out.)

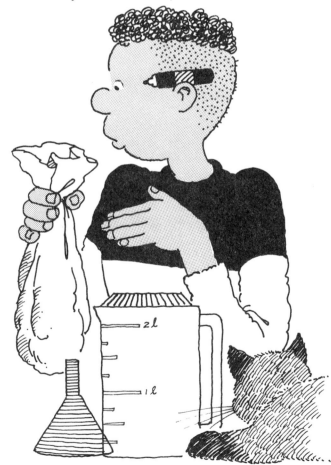

6. Close the bag tightly and hold it while you take it away from your mouth.
7. Slide your hand down the neck of the bag until the bag is completely expanded. Mark the bag at the point where you're holding it, in case your grip slips.

8. Push the neck of the funnel into the mouth of the bag, still keeping a firm hold on the bag so it doesn't move. Don't worry about the air escaping—you don't need it.

9. Using the marked container, carefully pour water into the bag until it's as fully expanded with water as it was with air. The bag will get quite heavy, so you may want to rest it on something as you fill it.

This will give you an approximate idea of your lung capacity.

The average 137 cm (4 foot 6 inch) tall boy has a lung capacity of approximately 2 L (2 quarts). The average 152 cm (5 foot) tall girl has a lung capacity of about 2.7 L (3 quarts).

Compare your lung capacity to that of friends who are the same sex and height as you are. If you have a greater than average lung capacity you have an advantage in endurance sports, like cross-country skiing, or long-distance running or swimming. For short, intense bursts of energy, like sprinting, large lung capacity probably isn't as important— many sprinters hardly breathe at all during the few seconds they're racing.

Lung capacity is something you are born with and cannot be expanded much except a little during your teens. But here's another breathing test you *can* improve on.

You'll need:
several friends

1. Start jogging around the block, keeping up the same speed.
2. Talk constantly as you jog. If you can't think of anything to say, recite nursery rhymes or repeat the alphabet.
3. Listen for who stops talking first.

The person who stops talking has probably "*run out of* breath." This has nothing to do with how much air your lungs can hold, but with how fast you can exchange the air in your lungs and how quickly and easily your heart can deliver the oxygen from the air to your muscles. Since both these things have to do with *muscle strength*—your chest muscles and your heart muscle—you can improve them with exercise. If you and your friends jog around the block regularly, soon none of you will stop talking while you run.

Try an Explosive Style
You'll need:
chalk
a wall you can make chalk marks on
a tape measure or metre stick

1. Put the arm you use most behind your back.
2. Rub chalk dust on the middle finger of your other hand.
3. Stand sideways against the wall, with your chalked hand next to the wall. Make a mark on the wall as high as you can reach. This is your base mark.
4. Put more chalk on your finger. Squat down and make a mark on the wall as high as you can.
5. Try it again, then measure the distance between your base mark and your best jumping mark.

6. To get your score, multiply the distance you jumped by your weight. How does your score compare to your friends'?

Explosive power gets you out of the starting block fast, up in the air to score the basket, spike the volleyball or defend the goal.

Look Through Your Binocular Vision

1. Hold this page with the line up against your nose, as in the drawing.
2. Focus both eyes on the circle.
3. What pattern does the line make?
4. Try the same experiment with the page held at arm's length.

What's this got to do with tennis? (Or baseball or. . .?)

The pattern you see tells you something about your binocular vision (the way your two eyes work together).

If you see an X crossing at the circle, you have good binocular vision. Tennis, baseball or cricket could be a cinch for you. Try the arm's length test to get a better idea of eye/ball distances involved in tennis.

If you see a V coming to a point before the circle, you'll think an approaching ball is closer than it really is and you'll have a tendency to hit early.

If you see a V coming to a point beyond the circle, you're more likely to think an approaching ball is farther away than it really is and you'll tend to hit late. You're unlikely to see this if you're really focussing both eyes on the circle. If this is what you see, maybe you should have your binocular vision checked.

39

COMING IN FOR A LANDING

HAVE you ever heard a cat land on its feet? Listen closely and what you *won't* hear is a sudden thump as all four feet hit the ground at once. Instead, you'll hear something like a short, soft drum roll—da, da, da, dum—as the cat hits the floor one foot at a time.

Cats may not understand science, but they know what to do with their momentum. Any object or body dropping through the air carries with it a certain amount of momentum. The bigger the object is and the faster it falls, the greater its momentum.

The only way you can stop dropping through the air is by having your momentum taken away from you. Robbing your momentum requires an opposing force, which is just what the ground supplies when you land on it. The ground actually pushes against you—if it didn't, you'd fall clear through to the other side of the earth.

The force that the ground exerts against a falling object is always sufficient to take away the object's momentum—to make the object slow down and stop. When you jump from an airplane, your momentum is huge, and the ground responds with an equally huge force.

Learning how to fall properly is a skill that many athletes work at diligently. Skateboarders practise tumbling off their boards so that they automatically fall safely. To avoid hitting the pavement with a painful thud, the skateboarder learns to roll off the board. Just like that cat landing one foot at a time, the rolling skateboard rider distributes the fall over an extended period of time. A rolling fall is like a series of short falls, rather than one big one.

Skateboarders also make sure to roll on the naturally padded parts of their bodies—the large muscle areas of their forearms, back, buttocks, thighs and calves. Parachutists do the same when they hit the ground. So do judo students, who learn to land with a roll when thrown.

If you watch baseball players catching a ball, particularly pro ball players who face balls coming at them at about 145 km/h (90 mph), you'll see them using a similar technique. As the ball strikes the mitt, the player pulls his arm back, gradually absorbing the ball's momentum. Try catching a ball the way ballplayers do, softening the catch by moving with the ball. Then try it again with your arm held rigid. (Don't do this with a real zinger!) Feel the difference?

How Gently Can You Land?
You'll need:

a styrofoam cup
water
a chair
a staircase

Trial No. 1
1. Stand on the **bottom step** of a staircase.
2. Leap off and land with your legs stiff.

Ouch! You've just won the disapproval of all your neighbourhood cats—you've landed the wrong way. Your body has absorbed the Earth's force all in one painful jolt. In athletic games, such as basketball or long jumping, this kind of landing can sometimes break bones or twist ankles.

Trial No. 2
Leap from the bottom stair again, but this time do it this way: make contact first with your toes, then the balls of your feet, and finally the heels. As your feet are making contact, bend your knees. How does it feel?

Using your feet and knees this way spreads your impact over a longer period of time, so you make a gradual acquaintance with the ground. It's like having a series of small forces hitting you one by one, instead of one big planet slamming you all at once.

Trial No. 3
So you think you're pretty good at landing now? Prove it. Fill the styrofoam cup with water and hold it when you jump off the bottom stair. The object is not to spill any water. Try bending your knees even more. Does bending your whole body help? Practise until you can do a dry landing. Make sure you wipe up any spills between jumps to avoid slipping.

Once you've accomplished a dry landing from the bottom stair, try jumping off a low chair without spilling water from your cup. (Don't land stiff-legged when you jump off a chair or you might hurt yourself.) When you succeed, tell the cats about it. They'll be impressed.

SAY "BASEBALL"

READ the title of this page out loud. That's about how long it takes for a big league fastball to move from the pitcher's hand to home plate—41/100 of a second. That's fast!

Many coaches and trainers say that if you want to be a top athlete, you've got to have a fast reaction time. You've got to be able to start to swing a bat or stop a puck or dive for the spiked ball in the blink of an eye.

How fast is your reaction time?

TRY THIS

You'll need:
a friend
a ruler

1. Hold one end of the ruler so that the other end is between your friend's thumb and middle finger, as shown in the picture. The ''1 inch'' mark should be right between her fingers.
2. Without warning, drop the ruler. Your friend must catch it between her two fingers.
3. Note the spot where your friend caught the ruler. Now you try it.

The lower the number at the point where you catch the ruler, the faster your reaction time, the time it takes for you to start moving.

Were you the fastest? The slowest? No matter how you ranked among your friends, as far as most sport goes, *it doesn't mean a thing*.

To understand why reaction speed alone isn't important, try the test again. This time, do it by yourself. Hold the ruler with one hand and try to catch it with the other. How did your reaction time improve so much in such a short time?

The answer is, it didn't. You probably still started to move at the same speed as before. But

this time you knew when the ruler was going to drop. Your ability to anticipate was more important than your ability to start moving quickly.

Similarly, most good athletes don't need to be superfast at reacting. Sprinters are an exception; they must react quickly to the start signal. In most cases it's what an athlete can predict about the course of the event that really counts. The champion boxer Muhammed Ali, famous for his speed and timing in a sport where both count a lot, had only an average reaction time. But he was way above average in his understanding of the sport and what his opponents were likely to do.

If you're a shortstop in baseball, for example, there's just no way you can merely react quickly enough to catch every ball that heads your way. Instead, you have to anticipate where the batter is going to hit the ball. When the ball does indeed head that way, and you're there to catch it, you look like a genius with lightning reflexes.

Squash is the same. It looks like a game that demands fast reaction time. But beginners and experienced players may have exactly the same reaction times. The difference is that beginning squash players have no idea where the ball is going to bounce so they scurry all over trying to keep up with it. But from hours and hours of time on the court, the experienced player knows exactly where the bouncing ball is going to be.

The better you are at predicting what will happen next in a sport, the less fast you have to be. If you're ready and waiting for the ball, or the puck or the punch, reaction time is of little importance.

RUNNING ARM IN ARM

Look at Fred and Ted. One of them is a sprinter, the other a marathoner. Can you tell which is which?

Here's a clue: sprinters must have extremely powerful legs to transport themselves as fast as possible across a short distance.

So if you guessed that Fred is the sprinter, you're right. But why are Fred's arms as muscular as his legs? After all, sprinters don't run on their hands!

Well, in a way they do. You run *on* your legs but you run *with* your arms.

Running or walking requires that the human body use its energy to move forward in a straight line, which is a lot trickier than you might think. Why tricky?

To find out, go for a stroll around the room. What does each arm do when the leg on the same side moves forward? What happens to your body if you hold your arms still?

Whenever you put a foot forward, you throw your body off balance. Swing your right foot forward, and your body automatically sways to the left. Swing your left foot forward, and your body sways to the right. Sway to the left, sway to the right, and you can end up weaving down the road like a drunk on a Saturday night.

People use their arms to counterbalance their leg action. Every leg movement is matched by an equal and opposite arm movement, which acts to minimize side-to-side swaying. Thanks to your arm motion, you can walk in a straight line.

Now think again about Fred the sprinter. Fred's legs have to be strong because he needs to reach top speed fast and maintain it for the short time of the race. But his legs are so powerful that they threaten to steer him way off course with every running step. So he has to have arms as powerful as his legs. When sprinters are in training, they not only practise running, they also spend hours beefing up their arms and upper body by pumping iron and playing games of medicine-ball catch.

A marathoner, on the other hand, needs endurance as much or more than speed. Ted doesn't have to have the same explosive leg power as Fred and he doesn't want to have a lot of extra muscle weight to carry for the hours he'll be running.

Try these goofy gaits

Here are some goofy gaits that mix up the natural leg-arm rhythm that keeps you moving straight ahead.

• Go for a walk, but reverse your regular arm motion. Whenever you swing your left leg forward, swing your left arm forward with it. Swing your right arm forward with your right leg. Keep this up as you move down the street. Don't be fazed when fellow pedestrians whisper, "Nerd."

• Walk at a regular pace but pump your arms back and forth twice as fast as you're moving your legs. You'll probably find that your legs will speed up to keep in time with your arms, proving once again the strong connnection between arm and leg action.

• Clasp your hands together and hold them against your chest. Run quickly. You'll probably find that your shoulders swing forward farther than they usually do, particularly as you go faster. Your shoulders are pinch-hitting for your absent arm motion.

• Get down on the carpet and do a goofy crawl. Put your right hand forward at the same time as you put your right knee forward. Now put your left hand and knee forward at the same time. Babies learn very early how to synchronize the movements of their arms and legs.

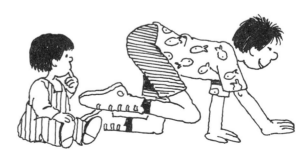

PUT YOUR FOOT IN IT

AVE you ever been walking along the street and suddenly tripped over nothing? Your friends look at you and sneer, "Walk much?"

What happened?

To find out, find an out-of-the way spot (preferably grass, because it's softer) where you can lie on your stomach and watch people walk by.

How high do they lift their feet off the ground?

Are you a daisy clipper?
When you walk through a field of daisies, do your feet clip the tops off the flowers?

That's what people used to look for when buying horses or dogs if they needed speed or endurance over long distances. An animal that barely lifts its feet off the ground (a daisy clipper) doesn't waste energy that instead can be used for doing the chores (hunting, racing, sheep herding, pulling a wagon). Today, there are more sophisticated ways of measuring how high an animal lifts its feet, but the best are still those who'd clip the heads off daisies.

The opposite of a daisy clip is a hackney stride. Perhaps you've seen horses that pick up their feet in a dainty, prancing gait. That's a hackney stride. It may look pretty but it's definitely not energy efficient.

Now, from your observations, do people make good daisy clippers? They should. Most pedestrians clear the ground by only about 1 cm (⅓ inch). Walking in this way is energy efficient—you don't do any more work than is necessary.

Unfortunately it doesn't take more than a little rise in the concrete for a daisy clipper to stub a toe. And then it's, "Walk much?"

Are you a floater?
Go back to your foot observation post and focus your attention on any joggers passing by. Can you tell the difference between the gait of a runner and a walker?

Runners "float." They spend part of the time with both feet off the ground. And the faster they run, the more time both feet are airborne. Walkers, on the other hand—er, foot—always have at least one foot on the ground. See if you can sense this difference as you compare yourself running and walking. Do you feel yourself "float"?

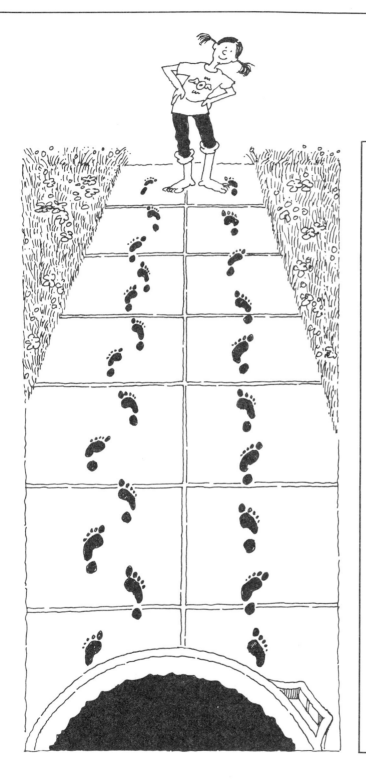

Footprint detective

Can you tell the difference between the tracks of someone running or walking?

Try This:
You'll need:
two bare feet
a sidewalk
a tub of water

1. Get your feet good and soaking wet.
2. Walk down the sidewalk. Look back at your tracks on the concrete.
3. Soak your feet again and run down the walk. Notice any difference between your walking and running footprints?

When you walk, your feet move parallel to each other, as if moving on two side-by-side tracks. When running, they follow one another, as if on a tightrope, making a single track.

Like daisy clipping, "single tracking" is an energy-saver. When your legs move on parallel tracks, you sway from side to side as you switch legs, which wastes energy. You have to use the parallel-track system when you walk in order to keep your balance. (Think how difficult it is to walk putting one foot directly in front of the other.) But it's easier to maintain your balance when you move faster, so you can manage with just one leg at a time right underneath the centre of your body. The energy you save by not swaying can be used to move your legs forward faster. To see how well it works, try running with your legs making parallel tracks.

ASK DR. FRANK

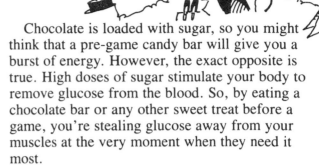

GOOD evening people in radioland and welcome again to *The Sport of Munching*. I'm Doctor Professor Frank Furter, and each week at this time I answer your questions about food and health. Tonight my topic is "Sports and Good Nutrition," and already I can see that my telephone hot line is heating up. So let's see who my first caller is. Hello, you're on the air.

Hi, Dr. Frank. My name's Charlene and I'd like to know if I can get extra energy by eating a candy bar just before I run in a track meet.

Charlene, think of your body as a car engine. It must be fueled regularly or else it will run down. The "gasoline" that you put into your "tank" comes mostly from food that is high in carbohydrates (the starches or sugars found in bread, fruits and vegetables, noodles, rice and potatoes).

Inside your digestive system, carbohydrates are broken down into glucose, which is a simple sugar that is shipped to your muscles to fuel their contractions.

Try this, Charlene. Pop an unsalted cracker into your mouth and chew on it for about a minute. Do you notice any change in the flavour? The mushy cracker should taste a little sweeter because the saliva in your mouth has begun to break down the carbohydrates in the cracker into sugar.

Chocolate is loaded with sugar, so you might think that a pre-game candy bar will give you a burst of energy. However, the exact opposite is true. High doses of sugar stimulate your body to remove glucose from the blood. So, by eating a chocolate bar or any other sweet treat before a game, you're stealing glucose away from your muscles at the very moment when they need it most.

Next caller please.

Hi, Dr. Frank, this is Bubba calling. The guys down at the health spa say that I have to pig out on protein if I want to build big muscles. Are they right?

Well, I could say it's all hog-wash but it is true that protein is the building block of muscle tissue. However, the amount of protein you need for building and repairing muscle tissue is not all that large, even if you're working out several times a week.

Unfortunately, many athletes think that they can have big muscles by gorging on protein. And so they eat lots of protein-rich foods—meat, milk and fish—and may even go so far as to consume expensive protein drinks and supplements.

The truth is, Bubba, that a well-balanced diet provides you with all the protein you need.

Next question.

Dr. Frank, my name is Julio and I play centre on my high school basketball team. My problem is that my mother always cooks me a couple of hamburgers and a plate of fries right before I leave for a game. Is it okay to play with so much food in my stomach?

Sure, Julio. After all, in basketball you're always trying to throw up, if you'll excuse the pun. Seriously, many athletes do put away a feast of meat and potatoes just before game time. But they'd be better off without it. A full stomach interferes with breathing. As well, the stomach muscles need extra blood to help them do the work of digesting; this diverts blood away from the muscles you use to play basketball.

Hamburgers and steaks, in particular, make a bad pre-game meal. Because they're high in protein and fat, they take a long time to digest. Tell your mother that you must avoid food for one hour before a game.

Next caller, please.

Dr. Frank, I'm Totie and I'm training to be a competitive figure skater, so I'd like to stay as slim as I can. What I want to know is, should I cut out fatty foods?

You'd be skating on thin ice if you did, Totie. Fat is an absolutely essential part of a well-rounded diet and is found in foods such as milk, cheese, nuts and meat. Certain vitamins—A, D, E and K—cannot be absorbed by the intestines without fat in the diet.

High-fat foods also provide energy. Carbohydrates are the main energy source in your diet but fats also serve as fuel, particularly in long-distance and endurance sports. So while you shouldn't eat a lot of fat, don't cut it completely out of your diet.

If you want to see for yourself how fat provides energy, I'll send you an experiment you can do at home. Thanks for calling.

Well listeners, I'm afraid that's all the time I have for tonight. Please tune in next week when *The Sport of Munching* will ask, "Does eating red vegetables help you to beet the odds?"

Totie's fat-finding experiment

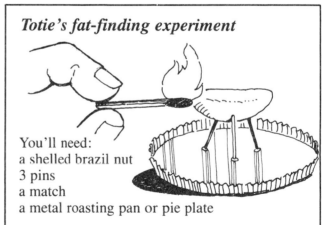

You'll need:
a shelled brazil nut
3 pins
a match
a metal roasting pan or pie plate

1. Ask permission to use the match.
2. Put the pan in the sink.
3. Stick the pins into the nut to make a tripod, as shown, and sit it in the pan.
4. Light the match and hold the flame to the nut until it catches fire. How long does the nut burn? Why does it keep burning so much longer than you think it would?

The fat in the nut provides the fuel for the fire, just as it would provide fuel for your muscles.

LOOK down on any baseball field just before a game and you might see a few players who look as though they're playing the game under water. They're slowly lobbing the ball back and forth or gently swinging the bat at imaginary balls.

These players are doing a slow-motion version of hitting and catching to start the blood flowing more quickly to the right muscles. This supplies the muscles with extra oxygen and raises their temperature. Warm muscles are stronger and have more endurance. A warmed-up baseball pitcher puts greater speed and distance into his throws, while warmed-up sprinters and marathoners get better running times.

Stretching exercises are another popular type of warm-up routine. Toe touches, hip rotations and other stretching activities loosen up the elastic parts of the body—the muscles, tendons and ligaments—and lower the risk of sports injuries. Football players, for example, are often fanatical about stretching, and more than a few wide receivers can rival any ballerina in doing the splits.

And by the way, human athletes aren't the only ones who warm up. If you've ever seen a horse race, you might have noticed the jockeys leading their horses slowly around the track or taking them for a gallop just prior to starting time. Horses are like people—they're better off warm than colt.

TRY THIS

The best exercises for any particular sport are those that use the motions of that sport. But there are many general warm-up exercises that are great for any sport. Here are some you can do any time you're planning to run, skate, throw, hop, skip or jump.

Hold yourself in each position while you count 15 seconds. Feel your muscles stretch, but not so much that it hurts.

Warm up 1
Bend forward and stretch down until you touch your toes. You can bend your knees a little, if that helps you reach your toes, then straighten them slowly once you get down there. Don't bounce!

Warm up 2
Steady yourself with one hand on a wall or chair back. Grab the back of your ankle with the other hand and bend your leg up until your heel touches your bum. Hold.

Warm up 3
Bending sideways at your waist, run hand as far down your leg as you can and hold.

Warm up 4
This is a good exercise to do before you run. Shift your weight to one foot, then raise the other foot so it's resting on the ball of the foot and rotate your ankle in one direction, then the other. Do the same with the other foot.

Weird warm up
Here's the warm up: squat like a frog, put your elbows inside your knees and push out. What's the sport? If you guessed basketball, congratulations. This warm up stretches the muscles you need for the sideways shifting that basketball players do.

ERE are some of the aches and pains you may get to know if you get involved in sports some day.

Biker's Bum

Ever had sore ischial tuberosities?

Cup your hands under your bum and sit on them. The ischial tuberosities are the two bumps at the bottom of your pelvis.

The constant rubbing of these bumps against a bike saddle can often lead to soreness. Numbness can also occur if the saddle pinches a nerve. Saddle sores and numbness are common though not especially serious complaints among cyclists.

Bikers often find themselves rubbing pavement the wrong way, too. ''Road rash'' is the all-purpose name for the scrapes and abrasions that follow when cyclist meets asphalt.

Turf Toe

What does a woman in high heels have in common with a football player?

They're both susceptible to turf toe.

It's especially a problem for football players, since landing on artificial turf can be like hitting pavement covered with only a thin layer of carpeting that doesn't ''give.''

When football players slam their big toe into one of these hard fake fields, the toe is often forced back into the foot, causing ligament and tissue damage in the joint. This injury, called turf toe, can be agonizingly painful.

Women who wear high heels sometimes suffer from turf toe, too. With all of a woman's weight bearing down heavily on her toes, stepping too hard can cause the very same kind of joint damage.

Swimmer's Shoulder

Swimming is actually one of the safest of all sports. It's a "low impact" sport, meaning that the body rarely hits anything that could harm it. Nevertheless, swimmers are prone to a few complaints, chief among them swimmer's shoulder, an inflammation of the shoulder tendons.

Tendons are fibrous pieces of tissue, somewhat like rope or cable, that connnect muscles to bones. They don't have a lot of give. Imagine rubbing a wire back and forth across a rough rock. Eventually, the wire gets warm. Similarly, a tendon that's overused gets inflamed. Swimmer's shoulder results when the tendons in the shoulder rub too hard against the top of the shoulder blade (the scapula) and the humerus (upper arm bone).

Swimmer's shoulder is just one form of tendinitis. Others include tennis elbow, Achilles' heel and jumper's knee.

Sprained Ankle

The sprained ankle is probably the most common injury among athletes. Sprains are injuries of the body's ligaments, which are pieces of tough, thick, rubbery tissue that connect bones to each other. Whenever a ligament is overstretched or torn, it's said to be sprained.

Swimmer's Ear

Have you ever wondered why there's wax in your ears? Swimmers often learn the answer to this question when they develop dizziness and nausea.

Chlorine is added to swimming pools to kill all the tiny "bugs"—bacteria and fungi—that breed in water and cause disease. Unfortunately, it also dissolves ear wax, in the same way that chlorine bleach removes grease stains from a dirty shirt. Since the job of wax is to protect the ears from bacteria and fungi, waxless ears are vulnerable to attack from these bugs encountered outside the pool and often develop middle ear infections. This throws off your balance and you feel dizzy and nauseated.

THE PAIN ALARM

WHOEVER coined the famous phrase, "No pain, no gain," had "no brain." Pain is Nature's alarm signal, warning you to stop whatever you're doing and seek help.

Sooner or later, if you play any game or sport, you'll run into some sports-related injury or medical problem. Cuts, breaks, sprains and strains are the painful partners of physical activity. Even though they're common, you should never treat injuries lightly.

Many top athletes are forced to cut short their career because they refuse to stop playing after an injury. Instead, they dope themselves up with painkillers and keep going. They forget that additional stress can turn a small, manageable injury into something serious and perhaps crippling.

So, whenever the pain alarm sounds, don't ask for painkillers and continue playing. Painkillers merely drown out the signal, they don't treat the injury.

Painkillers don't just come in bottles. Your body manufactures its own painkillers, chemicals released by your brain and spinal cord. These chemicals, which flood the body during physical stress, are called endorphins. They're also known as opiates because they're similar in many ways to the narcotic drug opium, and their painkilling effect is much the same.

Some athletes cope with pain by "running through it," letting their body take its own painkillers. They rely on a natural accumulation of endorphins during exercise to mask the pain. This strategy is really no different, and no better, than popping a handful of pills.

First Aid

The best response to an injury such as a sprain, twist, or bad bruise is "**RICE**," which doesn't mean order out for Chinese food. RICE stands for "**R**est, **I**ce, **C**ompression and **E**levation."

Rest. Stop as soon as you feel pain. If you continue, you could extend and possibly worsen the injury.

Ice. Apply ice to the wound. Ice limits swelling, which slows down the healing process. Don't apply ice directly to the wound or you could get freezer burn. Instead, put a cloth or towel over the injury first and then apply the ice. If you don't have ice, try a bag of frozen vegetables.

Compression. This also limits swelling. Gently wrap an elastic bandage around the wound, though not so tightly that you cut off circulation. If the injury is a sprained ankle, leave on the victim's shoe, and this, too, will prevent excess swelling.

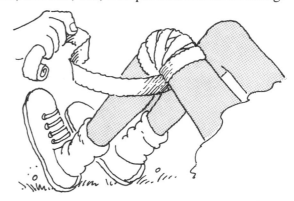

Elevation. If a leg or an arm has been damaged, put it up on something so that it is higher than chest level. Gravity will help in keeping too much fluid from accumulating in the wound.

These are only the initial steps in first aid. Always call a doctor to administer treatment after that.

HEN is a sheet like a football helmet?

TRY THIS

You'll need:

an egg
a sheet
two friends

1. Ask your friends to hold the sheet as in the drawing.
2. Stand back from the sheet and throw the egg at it. Throw the egg as hard as you can. What happens?

Why didn't the egg break? Because the sheet behaved like a football helmet. When a football player wearing a helmet is struck on the head, the helmet's padding absorbs and spreads out the shock. Your sheet did the same thing when the egg hit it. Compare this (but only in your imagination!) with what happens when you throw an egg against a wall, where the egg's shell has to take the full force of the impact. Now imagine that the egg is your head.

When you do that, it seems that only a dodo would ever fail to wear a helmet.

The same goes for face protectors. Puck sandwiches taste terrible, though that never used to stop hockey goal-tenders from dining on them regularly. And talk about fast food: these high-speed lunches would be delivered at up to 200 km/h (about 120 mph).

Eventually, goalies got smart and started wearing masks. Goalie Gerry Cheevers of the Boston Bruins used to decorate his mask with drawings of the stitches that he would have received if he had gone barefaced. The mask had more stitch marks on it than Frankenstein!

Though wearing proper protective equipment might seem like a common-sense idea, it's still just catching on. The batting helmets that baseball players wear can only protect them from fastballs moving less than 95 km/h (60 mph). And yet, most fastballs in the major leagues travel much faster than that. Many bicycle racers still don't wear helmets, even though they travel at speeds that could cause them serious head injuries if they fell.

What makes a good helmet? For starters, helmets should have an outer shell that stops penetration by sharp objects, such as hockey sticks or ski poles. So far, fibreglass has proven to be the most puncture-proof material. Also, chin straps on a helmet must be flexible enough to resist snapping when the helmet is struck. Finally, the helmet should have a liner like the sole of a running shoe, capable of absorbing or slowing down the strongest shocks.

Helmets have to match the sport they're used for. Contact sports, such as football and hockey, require helmets with padding that can bounce back after every blow, like a sponge. Crash helmets worn by racing car drivers, motorcyclists and downhill skiers have a different kind of lining, like styrofoam, which can absorb more energy but crushes on impact. After one strong blow, the helmet must be retired.

How do helmet makers know how much shock a helmet will absorb? By dropping it on its head. An artificial head, of course. The artificial head, wearing its helmet, is dropped from a height of about 3 m (3 yards) onto a padded iron block. Inside the head is a small instrument that records its speed during the fall and also how suddenly the head comes to a stop when it hits the block. A good helmet will absorb enough of the impact to let the head inside slow down before hitting.

This is important because the human brain is suspended in fluid inside the skull. Whenever the head is struck, the brain moves through this fluid and hits its hard casing. If the jolt is severe, the brain's nervous tissue and blood vessels may be badly damaged.

TRY THIS

Make a Helmet for an Egg
Conduct your own helmet test, but with an egg instead of a head inside.

You'll need:
a couple of eggs
a chair
a frying pan

1. Make a helmet for the egg. What makes a good helmet for an egg? That's for you to find out. You might want to wrap the egg in cotton batting, or maybe some newspaper. Or both. How about tying your helmet to a homemade parachute? The possibilities are endless.

2. Stand on a chair, hold the helmeted egg as high as you can and drop it into a frying pan sitting on the floor.
3. Make an omelette out of the non-survivors.

THE WIDE WORLD OF SPORTS

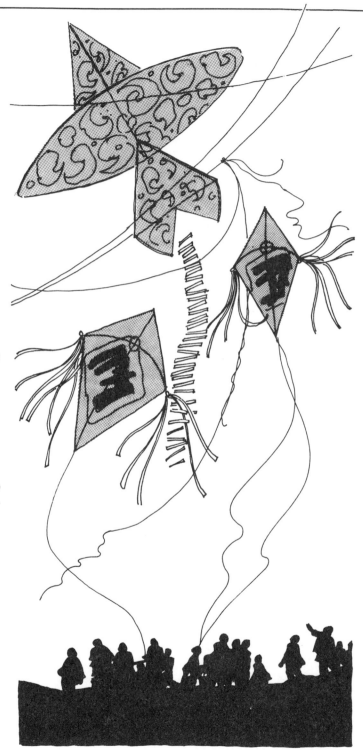

WHAT'S your favourite sport? Tossing the caber? Octopush? Sepak takraw? Never heard of any of them? You don't know what you're missing. There are hundreds of sports played around the world, and many of them are just as much fun as your own favourite sport.

Care for a game of...

Kite-Fighting

Are you a Charlie Brown kind of kite-flyer? You just get your kite up and soaring when it gets entangled with a kite-eating tree. Well, imagine *trying* to get your kite to tangle with another! That's a popular sport in India and Japan, where kites flying high above the ground duel in a vicious contest called kite-fighting.

If you were to enter an Indian kite-fighting contest, you'd find your kite connnected to two strings: one string to control the kite, the other string covered in powdered glass. The function of the glass-encrusted string is to cut through the strings of your opponent's kite and set it adrift.

Kite fighting is popular not only in India but also in some South American countries, where competition gets even more cutthroat. Razor blades are embedded in the framework of the kite, to sever your opponent's strings or rip holes in the kite.

In Thailand, kite-fighting competitions involve a ''male'' kite, or *chula*, which is shaped like a star and requires several men to launch it. Pitted against the *chula* is the *pakpao*, a team of two ''female'' kites, both diamond-shaped. *Pakpao* kites are only half the size of the *chula*, but what they lack in size, they make up for in speed and manoeuvrability. The *chula* is armed with bamboo slats with which it tries to entangle the tail of a *pakpao*. At the same time, a *pakpao* will try to loop a piece of string around one of the *chula*'s points and drag it down.

Snowsnake

Snakes dislike winter, unless, of course, they're Canadian snowsnakes. Made not from flesh and scales but from a young hickory or maple tree, the snowsnake is the main piece of equipment in an old woodland Indian winter sport that is still played today.

Many hours go into preparing the long, pole-like snowsnake—carving it out of a log, smoothing the surface and coating it with shellac, and finally, attaching an arrow-like head made of lead. There are two types of snakes: the one-metre ''Mudcat'' and the two-metre ''Long Snake.''

The object of the sport of snowsnake is to throw your snake down a long, narrow trough carved out of compacted snow. Any number of teams may enter the game, but each one is allowed only four throws. The winning team is the one that hurls its snowsnakes farthest down the track. If conditions are ideal, a snowsnake may travel up to a kilometre and a half (1 mile).

Snowsnakes can be launched overhand, like a javelin, or they can be thrown underhand or with a side-arm pitch. But watch out, eh! A well-tossed snowsnake is capable of slithering down the trough at up to 160 km/h (100 mph), and more than a few snowsnakes have been known to jump the track and take a sharp bite out of a nearby boot!

Octopush

Contrary to what you might expect, octopush is not a pushing match with an octopus. Introduced in South Africa in the 1960s, octopush is hockey played underwater. Players don skin-diving equipment, jump into a swimming pool, and with miniature hockey sticks and an ice-hockey puck, play by the normal rules of hockey on the pool bottom.

Of course, real hockey is always played on ice. Which is why the Canadian version of octopush dips below the surface of frozen ponds and lakes. Players outfitted in scuba gear enter the frigid water through a hole in the ice. Standing upside down with their fins against the ice, they play hockey with sticks and an inflated beach ball.

Canadian octopush is still largely an oddity, enjoyed by only a few. But who knows, every popular sport was once unusual and had only a few enthusiasts!

Sepak Takraw

Americans are crazy about baseball, Canadians hockey, Argentinians soccer, and Malaysians sepak takraw.

Sepak takraw is like volleyball, and it isn't. It's like volleyball in that two teams compete against each other across a net, and the ball must not hit the ground. It's unlike volleyball in that the ball is smaller and the top of the net is only about shoulder height. A takraw ball is made of woven rattan, has a hollow middle and is about the size of a softball.

But the real difference between volleyball and sepak takraw—and the real challenge to the sport—is that you can't hit the ball with your hands. Feet, knees, head, shoulders, elbows are okay, but no hands! It's said that keeping the takraw ball in the air calls for the speed of badminton, the dexterity of football and the teamwork of volleyball.

Tossing the Caber

Hoot maun, whit's yer caber? In the Heelands of Scotland yer caber looks a wee bit like a telephone pole, except that it's nae attached to the grun. It's a smooth, tapered tree trunk that guid kilted laddies toss durin' their annual Heeland Games.

The lad maks a platform wi' his hands and hulds the narrow end, leanin' the heavy pole agin' his shoulder. He then takes a run and heaves the pole into the air, aimin' for the perfect toss, wi' the caber landin' on its heid, then floppin' owr and smackin' the grun wi' its base pintin' awa from the competitor. It's nae how far the lad tosses that counts, it's the accuracy. A caber should pint in the same direction the lad was facin' when he throwed it.

TRY THIS

Sepak takraw has many rules but you might want to try a less complicated version of the game, called in-tossing takraw, which is played usually as a training exercise.

In-Tossing Takraw

You'll need:

a takraw ball, but if you can't find one, improvise. A takraw ball is 40 cm (16 inches) in circumference and weighs 200 g (7 ounces) . A ball of crumpled paper works well. Or try a hollow ball made of plastic, or a foam rubber ball with similar dimensions.

a group of players

The object of in-tossing takraw is to see how many times, and in how many different ways, a group of players can keep the takraw ball in the air without it touching the ground.

The game is most fun when you try more difficult or stylish shots. A simple kick is easiest, while shots with knees, elbows, heads or shoulders are harder. Try the classic takraw shot, which involves kicking the ball with both feet together behind the back. Remember: hands are not allowed in takraw!

One legendary takraw player, named Daeng, was said to have an infinite number of shots. His most spectacular feat, though, was dropping down on all fours and then rebounding from the ground, striking the ball with his rump on the way up. Can you do a ''Daeng'' shot?

RUBBER SOLE

Has the idea of competition ever given you cold feet? Well, this experiment will, too.

You'll need:
a pair of sneakers
a plastic bag
a freezer

1. Put your sneakers on and walk around in them. Rub your feet along the ground. Notice how much you slide. Bounce up and down.
2. Close your sneakers in a plastic bag and put them in the freezer. Make sure the plastic bag is a fairly thick one or you'll run the risk of having food that tastes like feet. Leave them there for 24 hours.
3. Put the sneakers on (they may be stiff so be careful not to tear them) and try your running, sliding and bouncing experiments again.

Why do frozen sneakers slide and bounce differently than warm ones?

Normally, there's a lot of friction between your sneaker soles and the ground. That's what allows you to get a good grip when you walk or run. Friction always occurs wherever two objects rub against each other. You can see why if you look at any object under a microscope. No matter how smooth the surface of that object may look to the naked eye, it will appear very bumpy or jagged under the microscope. Whenever two objects are pressed together, their uneven surfaces "interlock," like pieces of a jigsaw puzzle fitting together.

Sneakers give you a better grip on the ground because rubber is a soft material. It moulds itself more to the ground, providing more friction. Hard surfaces, such as the soles of leather shoes, make contact with the ground in only a few places.

When you freeze your sneakers, their rubber soles become harder. They no longer make as much contact with the ground so you slide where normally you'd grip.

In many sports, athletes want to increase friction to prevent sliding. Have you ever seen baseball pitchers rubbing a small bag between their hands? They're putting rosin (pronounced ''rozzin'') on their hands to increase friction and improve their grip on the ball. Pole vaulters use sticky adhesive tape on the pole or rosin on their hands to achieve a similar result. The covering on a ping-pong paddle improves the ''grip'' between the paddle and the ball, so you can give spin to the ball.

Mountain climbers use pressure to increase friction. A rock climber knows that by leaning well away from the rock face, he can thrust more firmly against the surface of the rock. This helps to hold the rock and soles of his boots more tightly together, and reduces the likelihood of slipping.

Of course, not all sports aim to increase the ''grip'' between two surfaces. Take bowling, for example. The best way to steer a bowling ball for a strike is to give it a spin so that it curves at the far end of the lane and scoots in behind the head pin.

Unfortunately, because bowling lanes are so long, a ball with a spin on it will quickly veer off into the gutter. That's why the first 10 m (30 feet) or so of a bowling lane is always covered with a light coating of oil. As the bowling ball rumbles over this oil slick, it spins but it doesn't curve. Only when the oil slick thins out and the ball makes contact with the wood surface of the lane does it finally start to swerve. Friction between the ball and the wood acts as a sideways force, making the spinning ball curve sideways and, with luck, into position for a strike.

SWEET SPOT

I F someone told you that a baseball bat had a sweet spot, would you say, "Make mine chocolate"?

Well, don't get too hungry because what makes this spot sweet is the love ballplayers have for it—it's the point on the bat that makes the ball travel farthest.

How do you find it? Not by tasting the wood, but by applying this test.

You'll need:
a baseball bat
a baseball

1. Hold a wooden bat near the handle as shown between the thumb and index finger of one hand. If you have an aluminum bat, hold it about one-quarter of the distance from the top.
2. Take the baseball in your other hand and tap it against the bat's handle just below where you're holding it. You'll feel the wood vibrating in your hand. (An aluminum bat not only vibrates, it hums!)
3. Continue tapping as you move down the barrel to the fat end. Somewhere along the bat you'll find a spot where your tapping won't cause the bat to vibrate or move in your hand. You've found the sweet spot!

Many things vibrate when hit. However, every vibrating object has one or two places where the vibrations are very small and are hardly noticeable. These quiet areas are called nodes. The sweet spot is near a node.

Once you've found the sweet spot on your bat, mark it and then have a friend pitch you a few balls. You'll find that hitting a ball with the sweet spot feels better. Your hit will seem effortless, and your hand won't sting the way it does when you hit a ball with other parts of the bat. It will even have a satisfying sound—a clear and solid "crack."

Why does the sweet spot feel so sweet?

You'll need:
a baseball
a bat
a friend

1. Crouch down and hold the baseball bat out in front of you.
2. Have your friend drop the baseball so that it hits the bat at various points. Watch how high the ball bounces each time. At what point on the bat does the ball bounce highest?

A moving bat and a moving ball each carry with them a lot of energy. When the two collide, the question always is: will their energy combine to launch the ball as far as possible into the park, or will some of it be wasted? Hitting a ball with the bat's sweet spot puts most of the energy into giving the ball a good, sound launch. Hitting anywhere else on the bat wastes some energy by transferring the energy into vibrations and movement of the bat.

Baseball bats aren't the only pieces of sports equipment with sweet spots. Tennis rackets, ping-pong paddles and even golf clubs have them. Can you find them?

Sweet spots are sometimes called joy spots. You may have heard that sports equipment manufacturers make it easy to find the sweet spot by sticking the label on top of it. Sometimes this is true, sometimes not. The only reliable way to find the sweet spot is to use the above test.

The Care and Feeding of Bats

- Don't pound home plate or the ground with your bat.
- Don't leave your bat out in the rain. Moisture warps the wood and raises the grain.
- If the grain on your bat is raised, rub it down with a smooth bone or a piece of hard wood. This is called burnishing.
- Rub oil into the bat whenever it gets wet, and also before putting it away for the winter.
- Store your bat in a cool, dry place and keep it in a vertical position. If possible, hang the bat from the beams in a dry cellar.

WALKING IN SOMEONE ELSE'S SHOES

 OU'LL need:
a friend whose feet are the same size as yours
her well-used shoes

1. Put on your friend's shoes and have her put on yours.
2. Go for a walk.

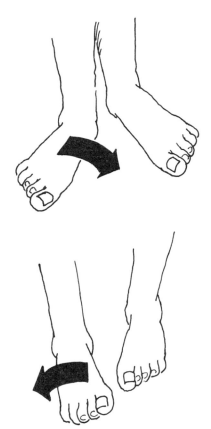

Feel awkward? Find your feet moving a little differently than usual? That's because well-worn shoes are the leather memory of their owner's gait. Put them on and they force you to walk the way she does. And everyone has a slightly different gait.

A normal footstep advances smoothly, beginning at the heel and rolling forward to a firm lift-off at the toes. Your foot comes down on the outer edge of your heel, then rolls forward on what would be the skinny part of your footprint, then inward to the ball of your foot.

If your foot has a tendency to roll farther inward toward your big toe as it moves to push off, this is called pronation. If your foot rolls outward as it pushes off, you supinate. Most people pronate slightly and this is why the average person walks with slightly turned-out feet.

But some people have really exaggerated rolls—they **overpronate** or **oversupinate**. Both types of people look a little like birds: overpronators walk in a duck gait; oversupinators are pigeon-toed.

But overpronating or oversupinating shouldn't get in the way of your athletic success. Many famous athletes overpronate or oversupinate. For instance, Bob Hayes, who was once the world's fastest runner, had a pigeon-toed gait.

There are disadvantages, though. Sometimes pigeon-toed runners trip over their own feet, or they may develop tendinitis or ankle sprains. Like-wise, duck walkers can sometimes develop shin splints and knee and hip problems.

In that case, the sole-ution calls for exercises to tighten up foot muscles or even for special ortho-paedic shoes.

TRY THIS
Play Sherlock Holmes with Your Shoes

How do you walk?

Turn your shoes over and examine the wear and tear on your soles. If you've worn down the out-side rim of your soles, you're a supinator. If you've worn down the inside of your soles you're a pronator. Or maybe you fall (or should we say walk?) somewhere in between.

AS THE FOOTBALL TURNS

HAVE you ever seen a seal balance a ball on its nose? It looks easy. . .until you try it yourself. If your nose knows it isn't having a ball, try to balance the ball on your finger, the way a basketball player does. Sit the ball there and it falls off. But twirl it the way the basketball player (or the seal) does, and you can keep it perched on your finger for much longer.

The secret is all in the spin. The faster the spin, the better the ball behaves.

Lots of ball sports use spinning balls. Football players spin the ball for greater control as they throw passes. A well-thrown short pass is sometimes called a ''bullet pass,'' and for good reason. Bullets also spin. A rifle barrel has spiral grooves carved along the inside of its barrel that put a spin on the bullet as it passes through.

What happens without the spin? Try throwing a football without a spin. The result is a quarterback's nightmare. Tumbling head over heels, the ball will veer off course and plop to the ground in a hurry.

Frisbees spin as they travel, too. How far will a frisbee go without the spin?

Spinning footballs, bullets and frisbees all benefit from gyroscopic stability. This is the tendency for a spinning object to keep its axis (the centre around which it's turning—think of the hub or axle of a bike wheel) pointed in a constant direction.

Does that mean a spinning football should fly straight to its target? Not really. As the ball travels through the air, it loses some of its gyroscopic stability to air resistance. This creates a ''wobble,'' called precession. The football no longer points straight ahead. Instead, its ends turn like a corkscrew, drawing small circles in the air as it moves.

Any spinning object that is not perfectly balanced will precess—even our own Earth. The Earth is spinning around its own axis, making one revolution every 24 hours. The Earth's spin is not perfect, however, due to the gravitational pull of the Sun and Moon. So the Earth has a wobble. A very slow wobble. Scientists estimate that it takes about 26 000 years for the Earth to precess just once.

TRY THIS

Turn a magarine top into a margarine top

You'll need:

a couple of ballpoint pens

a couple of plastic lids (e.g., from margarine and yoghurt containers)

a large pad of paper

1. Punch the ballpoint of a pen through the centre of the margarine lid so that the tip of the pen pokes a couple of centimetres (half an inch or so) through the other side.
2. Press the ballpoint against the paper. Twirl the end of the pen in your fingers or between your palms, as if you were twirling a top, then let go.
3. Watch it draw. (If the ballpoint pen doesn't work after several tries, use a sharp felt-tipped pen instead.)

How can you get your pen top to draw patterns? By allowing it to precess across the paper. A tilted top will precess as gravity tries to pull it over. Gradually, you'll see the loops increase. The large loops are precession drawings.

Want to add small loops for prettier patterns? Try adding bits of Plasticine to one edge of the lid. You may find small loops in your drawing even without the added Plasticine—that's because the pen is a little off-centre in the lid.

Use other lids and pens for home-made tops of different sizes. Try poking the pen well off-centre through the lid and see what the top draws.

THE WHEEL STUFF

WHICH is faster, arms or legs? That may sound like a funny question. After all, who races with their arms?

Wheelchair athletes do.

But can wheelchair racers push as fast as legged athletes can stride? You bet! In a marathon race between a top runner and a top wheelchair athlete, the competitor in the wheelchair would not only win, he'd have time for a shower and a cold drink before his opponent reached the finish line.

Giving 'Em the Gears

Everyone is familiar with the gears on a bike, but not many people realize that a wheelchair has ''gears,'' too.

On a bike, changing gears means flipping a lever so that your bike chain changes to a different size of ring. On a wheelchair, you change gears by moving your hands between the outer and inner rims. Both produce the same result. Gripping the smaller, inner rim gives you more power per push: the wheel travels farther for the distance your hand has to move, but it takes more effort for you to move it. If you push the small rim while you're moving, you get more speed, but it's very hard to get started from a standstill this way. On the other hand, gripping the large outer rim makes it easier to move the wheel, but your hand has to travel a lot farther compared to how far the wheel goes. So the outer rim is your better ''starting gear.''

Shooting from the Hip

When a standing athlete tosses a basketball into the basket, she uses her whole body to propel it. A wheelchair athlete may have muscle control only in her arms and shoulders. This adds to the skill required to get the basketball into the basket.

TRY THIS

You'll need:
a kitchen chair
a basketball
a hoop (or pick a point on an outside wall that's about hoop height)

1. Position the chair so it faces your hoop.
2. Keeping your body in contact with the chair, try to throw the basketball into the hoop.

Do you use different muscles than you would when throwing from a standing position?

THAT'S THE WAY THE BALL BOUNCES

I N the major sports leagues, it's not just the athletes who have to stay in top condition. Sports equipment must be kept in good shape, too, or risk being thrown out of the game.

How do you check if a ball is in good shape? By the way it bounces back!

In every major ball sport, there are standards for measuring a ball's bounciness. And balls have to meet those standards rigidly. . .er, bouncily. Why? Think about basketball for example. If every ball had a different bounce, the play would change every time a new ball was used. Imagine dribbling a basketball with the bounciness of a superball then switching to a basketball that bounced like a baseball. The same goes for all other ball sports.

Balls bounce because when they hit the floor they are flattened out. Things made of rubber tend to return to their original shape after they've been squashed, and a ball is no exception. It rebounds from the floor as it restores itself to its usual round shape. The bounciness tells you how quickly the ball is restoring itself to its former roundness. When a ball can no longer restore itself well, it's bounced from the game. In general, a ball lasts one year in a major sports league.

What kind of standards do balls have to meet?

When dropped from a height of 182.9 cm (72″), a basketball must rebound to a height of between 125.5 and 137 cm (49 -54″).

When dropped from a height of 254 cm (100″), a volleyball or soccer ball must rebound to a height of between 152.4 and 165.1 cm (60-65″). These figures are set out by the sports leagues. Balls for home use are made less expensively than those made for the professional leagues, so their bounciness will be different.

TRY THIS

How bouncy are the balls in your house? Check them out with this test.

You'll need:
a variety of balls—for example, a rubber bail, superball, tennis ball, basketball, volleyball, soccer ball
a metrestick (yardstick)
a pencil or felt pen
a hard floor
some notepaper

1. Hold the first ball at the top of the metrestick and drop it.
2. Mark the measuring stick at the highest point of the ball's bounce. (It may take a couple of tries until you get practised at noticing where the top of the bounce is.) Which ball bounces highest? Note your results.
3. Try the test again on different surfaces. Which produces more bounce, a vinyl floor or a wood floor? How about a fresh-cut lawn, concrete sidewalk, mattress or gravel road? Do different surfaces affect the bounce ranking of your ball collection?

Thermo-Bounce

A strange argument raged during the Chicago White Sox/Detroit Tigers weekend baseball series in Chicago in July 1965. The Tigers accused the Sox of illegally refrigerating the baseballs. They said that was why only 17 runs had been scored in five games.

"Ridiculous!" answered the Chicago team, pitching an accusation in return. They said that during the previous five games in Detroit, the balls played as though they'd been cooked! How else to explain the 59 runs scored, including 19 home runs!

Were these players crazy? Or do hot balls really have more bounce than normal, while cold balls have less?

You be the umpire. We'll probably never know whether the two teams really did tamper with the balls. But you can find out if there was a basis for all their name-calling.

TRY THIS

You'll need:
the same balls and equipment you used for your bounce test, and the results of your testing
a freezer
an oven

1. Chill the balls in a freezer for one hour.
2. Bring them out one at a time and test the bounciness of each ball as soon as you take it from the freezer. Note the results.
3. Wait a few hours to make sure the balls have thoroughly thawed, then sit them on a rack or cookie sheet and pop them in an oven at 105° C (225° F) for 15 minutes.

Make sure none of the balls is close to the oven elements.
4. Give them the one-at-a-time drop test again and note the results.

Were the baseball teams right? Could they really cook the results by cooking the balls?

HOW THE GOLF BALL GOT ITS DIMPLES

ALONG time ago, golf balls were called featheries. You made one by stuffing boiled goose feathers inside a casing of untanned bull's hide. You then sewed the feather-stuffed leather shut, moulded it into a round shape and painted it white. When you hit it, you hoped the feathers would help it fly. And it did fly better than the previous type of golf ball, which was carved from boxwood.

The featheries flew for 150 years. Then, in the mid-1800s, a clergyman at St. Andrews University in Scotland fashioned a new type of golf ball out of gutta-percha, a rubbery substance from India.

Excited by his invention, the clergyman took his gutta-percha ball, nicknamed a "guttie," out to the course. His first hit was a failure. The guttie sank sharply back to earth after flying only a short distance. The clergyman didn't give up easily; he hit the ball again and again and eventually noticed that as his club pitted and scarred the guttie, it stayed airborne longer.

Soon, golf ball manufacturers were mass-producing the guttie with specially moulded craters, or "dimples," on its surface. The more dimples they added to a ball, the farther it sailed. Today, golf balls, which are made of rubber with a hard enamel coating, have more dimples than Shirley Temple.

Why are dimpled balls superior to smooth ones? The answer lies in the way the air flows around the ball as it moves.

As the ball moves, it drags along a very thin layer of air, called the boundary layer. On a smooth ball, the boundary layer breaks away before it gets completely around the ball, leaving a very large "wake" dragging behind the ball like a

boundary layer separates

delayed separation of boundary layer

parachute. This puts a brake on the ball's momentum and sends it into an early nosedive.

To get rid of that parachute, you have to make the boundary layer stick to the ball all the way around to the back. Dimples mix the boundary air with the next outer layer of faster-moving air, giving the boundary layer an extra push that carries it to the back of the ball. The wake behind a dimpled ball is much narrower, and the ball sails farther.

On average, dimpled balls travel about four times the distance of smooth ones.

Putting around

- Probably the most unusual golf game ever was played on the moon by Captain Alan Shepard in February 1971.

- Floyd Satterlee Rood played the longest golf game ever. He putted from the Atlantic coast to the Pacific coast of the United States during a game that lasted nearly 13 months.

- The youngest player ever to shoot a hole in one, according to the *Guinness Book of World Records*, was Coby Orr of Littleton, Colorado. He was only five years old at the time.

SPORT MYSTERY 2:
WHY DON'T SPINNING SKATERS GET DIZZY?

RICK question! Spinning skaters *do* get dizzy—but they use their eyes to keep from losing their balance. Which is a little odd because twirling skaters (and everyone else) get dizzy because of their ears.

Inside each of your ears, past where you can reach when you wash, are fluid-filled canals. In the canals are jelly-like capsules sprouting tiny hairs that are actually sensors that send messages to the brain. When your head turns, the fluid moves in the same direction, but lagging a little behind at the start. As it lags, it presses against the hairy capsules, triggering the hairs to transmit a message about the direction and speed of your movement.

After you've been spinning for a little while, the fluid catches up with the canal and you have to rely on messages from your eyes and your muscles to tell you that you're still turning. When you suddenly stop, you feel like you're still moving because the fluid in your inner ear hasn't stopped moving yet. You also feel as though you're moving in the opposite direction, because the still-moving fluid is pushing against the other side of the jelly capsule.

Now your brain is getting contradictory messages—your muscles say you've been going one way and you've stopped, while your ears say you're going the other way and you're still moving. Your eyes are no help because they still haven't focussed on anything. This confusion is what we call being dizzy.

How do spinning skaters use their eyes to keep from falling over from dizziness? With practice, skaters learn how to focus their eyes intently on a stationary object as soon as they stop spinning so their brain can sort out the mixed messages more quickly.

TRY THIS

You can see the way the fluid in your ear canals lags behind the movement of your head.

You'll need:
breakfast—some milk and floating cereal (o-shaped oats are good) in a bowl

1. Gently spin the bowl in one direction. Watch the cereal to see when the milk it's floating on starts moving.
2. Stop the bowl. Watch which way the cereal goes. (Over the edge doesn't count.)
3. Eat the cereal.

THE SKIER'S AMAZING ONE-SECOND WEIGHT LOSS PROGRAM

OULD you like to lose some weight in a real hurry?

You'll need:
a set of bathroom scales

1. Stand up straight on the bathroom scales and take note of how much you weigh.
2. Lower yourself quickly by bending at the knees. What happens to the reading of your weight?

What you just did on the bathroom scales was **unweight** yourself. Unweighting is a technique that downhill skiers use to do turns. Unweighting requires you to bend your knees just before entering the turn. This lifts your body's weight off the skis for a brief moment, making it easier to turn the skis in a new direction.

Why do you lose weight so suddenly? When you bend your knees suddenly, you remove the support from under your upper body, which leaves it falling through space and only your feet pressing down on the skis (or bathroom scale). If you are quick enough, even the weight of your feet will be absent.

Unweighting lasts for only the split second that your body is falling through the air. When you stop bending your knees, the force of your body hitting the skis is so great that you suddenly "gain" a lot of extra weight—you blimp out for a moment.

Even if you've never skied, unweighting is probably familiar to you, though you may not have recognized it as such. Remember the last time you were walking barefooted and stepped on a sharp object? What was your immediate response? You probably bent your knee, dropped your hips and rushed your other foot forward to support your weight. In doing this, you reduced the force holding your foot against the sharp object—you unweighted your foot.

77

WHAT A DRAG!

Do you think sports are a drag? If so, you're right.

Drag is the resistance that a skater, skier, bicyclist or any other fast-moving athlete, encounters when he or she moves through air (or water). Drag opposes motion.

Skiing Eggs

Downhill skiers aren't being chicken when they crouch with their chests to their knees as they ski. They're doing the "egg." This skiing position was developed after putting skiers in a wind tunnel and having them scramble from one position to another while scientists measured air resistance.

These hard-boiled tests revealed that the "egg" is the best shape for downhill racing. They also found that protruding buckles on a skier's boots can add 0.3 seconds to every minute of a racer's time because they increase drag! That 0.3 is a mighty big number when skiing victories are measured in hundredths of a second.

Bent Skaters

Speed skaters have found that bending over with only their head and shoulders facing into the wind cuts down the amount of surface area hitting the air and is the best way to reduce drag. (The arm that swings back and forth does so to maintain the skater's balance.)

Drag-Free Clothes

Most sports clothing is definitely not a drag. The smoother the material, the less drag there is. Think of the shiny, form-fitting suits the speed skaters wear. The same applies to water sports, too. Competitive swimmers even shave their bodies to make their skin smoother and freer from drag.

On the other hand, you can go too far in combatting drag. Not long ago, one ski wear designer made a plastic fabric so smooth that skiers who fell couldn't stop themselves from sliding down the mountain. The plastic skin was so slippery that there wasn't enough friction between the plastic and the snow to slow down the skier's slide.

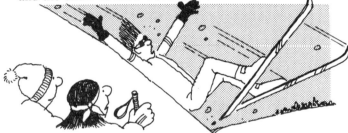

Drag is an obstacle in many sports, and the faster the sport, the greater the drag. If you increase your speed by ten times, your drag could increase by one hundred times! You can probably think of lots of sports where overcoming drag is important. What about horse racing, where jockeys curl over the horse in their own version of the egg? And how about drag racing?

But can you think of a sport in which competitors look at drag as a blessing? You're right if you guessed skydiving. If you were skydiving, you'd want to slow down the rate at which you moved through the air. Without a parachute on your back to pop open and give you some much-needed drag, your meeting with the ground would be a real drag.

TRY THIS

You don't need to be an athlete to experience the effects of drag.

You'll need:
a bicycle
a hill away from traffic
a stopwatch or a watch with a second hand

1. With just a push off, coast down the hill, sitting straight up as if you were driving a car. Clock yourself and note the time.
2. Do it again, but this time crouch down into the egg position. Clock yourself again and compare the results. Was there a difference? Why?

If you don't have a bicycle

You'll need:
a spoon
a tub full of water
sifted flour

1. Sprinkle the sifted flour on top of the water until it forms a thin, uniform layer.
2. Run the bowl of the spoon through the surface of the water. Feel the resistance in the handle of the spoon? That's drag. The turbulence patterns you see in the water are caused by the energy used in overcoming drag.

BACK in the 11th century, a man called the Saracen of Constantinople decided to try to fly. He fitted wooden slats into a flowing robe and, flapping these makeshift wings, leapt from a high tower. His flight was a short one. It ended with a "splat!" on the ground below.

As time went on, would-be birdmen from all over Europe continued "hang gliding" off castles and cathedrals. They all contacted the ground with the same rude jolt.

A few inventors had some success with hang gliders in the 19th century. But it wasn't really until the 1960s that hang gliding caught on as a safe and popular sport, thanks to NASA's idea for a kite-shaped parachute that would move forward through the air instead of just settling down to Earth. Although NASA scientists dropped the idea, California dreamers built their own modified versions of the kite. Strapping themselves in, they launched themselves off Pacific oceanside cliffs, Arizona sand dunes and the peaks of Vermont and into a new sport.

80

TRY THIS

Can you make a piece of paper *lift* without touching it? If you can, you'll have an idea of how *hang gliders* (and other aircraft) stay up.

You'll need:
a piece of paper

1. Hold one edge of the paper between your thumb and your index finger, letting the rest of the paper droop over your other fingers, as in the drawing.
2. Now bring your mouth close to your thumb and blow hard over the top of the paper. What happens? You've created an essential ingredient of flight, lift.

What's lift? Whenever an aircraft wing moves into the wind, it cuts the airflow in two. Instead of one *airstream*, there are now two, one flowing over the top of the wing, the other flowing under the wing. If the wing had been built with a curved top and a flat bottom, the airstream running over the top follows a different path than the airstream passing under the bottom. This creates a difference in air pressure between top and bottom of the wing and ''lift'' is the result.

By blowing over your piece of paper, you provided the airstream that lifted it.

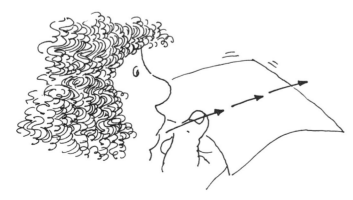

Playing the angles

Frisbees need a lift, too. That's why they're slightly curved on top, to break the airflow into different paths. But the curve on a frisbee isn't enough to take it far.

Your throw makes a difference.

Try throwing a frisbee parallel to the ground. Watch how high it goes and where it lands. Go back to your starting point and throw again, this time tilting the frisbee so the front edge is slightly higher than the back edge. Throw it with a tilt. What's different about its flight path? Can you find the angle of tilt that makes it go farthest?

TURN, TURN, TURN

WHICH of these figure skaters do you think would twirl the fastest?

If just thinking about it makes you dizzy, try tracing the figures onto cardboard and cutting them out. Then stick a pin where the dot is and try spinning them. Give them each the same amount of "push". Do you notice a difference? Is one harder to spin than the other? It may be hard to tell on the models, but in real life, the sit-spinning skater would twirl much faster and more easily than the one with her arms stretched out.

What would happen if a skater changed positions while spinning? If you have a stool or chair that spins, you can find out.

TRY THIS

1. Sit on the stool or chair with your arms outstretched.
2. Spin yourself or, even better, get someone else to spin you.
3. As you're spinning, fold your arms in tightly against your body.

Why do you and the figure skaters spin more slowly when you're stretched out than when you're tucked in?

A spinning object has angular momentum, a special property which is a product of the object's mass, radius (how far it extends from the centre of rotation) and rate of spin.

One of the things that makes angular momentum so special is that it always stays the same. This means that if you increase your radius while spinning (by stretching your arms out), either your rate of spin or your mass must decrease to keep the angular momentum the same. Since you're not likely to lose any mass (unless you go on a crash diet while spinning!) your rate of spin slows down instead. If you make your radius smaller by bringing your arms in, your rate of spin speeds up again.

SPORT MYSTERY 3:
THE GREAT BOARD-BREAKING MYSTERY

DID you ever get really angry and smash your hand down on a table? Hurts, doesn't it? And you didn't even get the satisfaction of denting the table!

If all you could manage was a sore hand, a solid table and a stupid look, how do karate experts break boards without breaking their hands?

Part of the secret lies in the fact that the board bends when the hand comes down on it. As the board bends, its upper half compresses, or squeezes together, while its lower half experiences tension, a kind of stretching apart. As it stretches, the lower half of the board starts to crack. The crack quickly spreads upward and the board breaks in two. If you look at the way boards are set up for this demonstration, you'll notice they are usually supported only at the ends, which gives them lots of room to bend.

Another part of the secret lies in the karate expert's aim and speed. The hand is moving at full speed when it hits the board because it's aimed at a point below the surface.

Why doesn't the hand break, too? Because some of the stress is absorbed by the skin and muscles lying between the bones and the wood. Also, some of the force is transmitted to other parts of the body. Karate experts are careful to hold their hand in certain positions, called "knife hand" or "hammer fist," and make contact only with the portion that can absorb the stress best.

Scientists estimate that in a karate "chop," the human hand can apply nearly six times the force it takes to break a pine board and nearly a third more force than is needed to break a concrete block.

So your hands are a lot tougher than you thought they were. But don't assume that this makes you an instant expert. A karate student studies and practises for years before she dares to bring down her powerful knife hand or hammer fist on a block of wood or slab of concrete. Doing this without instruction and training can be very dangerous.

WATER, WATER EVERYWHERE

TRY THIS

You'll need:
an ice-cube
a medium or large paper-clip

1. Open the paper-clip so you have a relatively straight piece of wire.
2. Put the ice-cube on a dish or paper towel.
3. Place the straightened paper-clip across the ice-cube. Hold on to the ends and press down hard for several minutes.

A H, there's nothing like a cold winter day when you can zoom downhill on your water skis. WATER SKIS?????

Yup. When you get down to it, every skier is a water skier. Downhill and cross-country skiers may think they're gliding on snow, but really they're slipping over a thin film of water. When skis rub against the snow they create friction, which melts the snow into water. Without this layer of moisture, snow skis couldn't slide—they'd be snowbound.

You've probably seen curlers furiously sweeping the ice with a broom. It's not because the ice is dusty. Sweeping the bristles against the ice also produces a thin film of water on which the curling rock slides. Toboggans, bobsleds and sleighs—they, too, glide over water.

Ice skates also travel across a film of water. The friction caused as you glide melts the ice slightly. But there's a little more to ice skating than just friction. Ice skating is a bit of a high pressure sport. Skate blades are so thin that your weight is concentrated on a very small area of the ice. This pressure alone can melt ice when it is not far below freezing.

The pressure you're putting on the ice-cube creates a layer of water, just as it does under your skates. Because the melted ice-water is still very cold, as soon as the paper-clip passes through it, it starts to refreeze. Although this gets tricky at room temperature, once the paper-clip is a couple of paper-clip widths deep into the ice-cube, there should be enough refrozen ice over it to let you pick up the cube just by lifting the paper-clip. This could inspire you to present. . . .

The Amazing Wire Through Ice Illusion

You'll need:

a plastic juice pitcher
1 m (1 yard) picture wire
2 bricks or other similar weights
winter: this experiment works best when the temperature is between 0°C (32°F) and −10°C (14°F).

1. Fill the pitcher almost to the top with water and put it outside (If the outside temperature is close to 0°C/32°F, it would be better to put the pitcher of water in the freezer.)
2. Tie a brick to each end of the picture wire.
3. When the water in the pitcher is frozen, carefully remove it from the pitcher and take it outside. Stand your ice block on a railing or a piece of wood or a couple of bricks so that when the picture wire is laid across the top of the ice block, the bricks tied to either end hang straight down.

4. Leave it there until the wire works its way about half-way down the ice block. (This could take a day or more depending on the temperature outside.)
5. Challenge your friends to figure out how the wire got into the middle of the solid block of ice.

Sport mini-mystery:
Why do cross-country skiers wax their skis?

It's not to make them slippery. In fact, it's just the opposite—it's to make them grip the snow. That's the technique of cross-country skiing: first, the right ski grips the snow while the left ski glides forward, then the left ski grips the snow while the right ski glides forward.

When you step down on a cross-country ski, the snow and the wax on the bottom of the ski actually lock together. The little points, or "arms," on the snowflakes stick into the wax and and hold the ski in place. This connection between the wax and the snow is just strong enough to provide a base from which to push yourself forward.

Different kinds of snowflakes require different kinds of wax. Newly fallen flakes are like the ones you see on Christmas cards—they have long, pointy arms that can grab onto a hard wax. But as snowflakes grow older, their arms grow shorter. Middle-aged snowflakes with their shorter, rounded arms require a soft wax. Eventually snowflakes have no arms at all. They can only grab a wax so soft and gooey it comes packaged in a squeeze tube, like toothpaste.

DO YOU FEEL A DRAFT?

IMAGINE what it would be like to pedal your bike ten times faster than normal—as fast as a dragster. A *bicycle* really did go that fast a couple of years ago on the Bonneville Speedway in northwestern Utah.

To start out, the bicycle and its rider were towed behind a high-powered car. Attached to the rear of the car was a large rectangular windscreen, which formed a wall between the car and the bike. When the car reached about 100 km/h (60 mph), the tow chain was cut, setting the cyclist free to pedal on his own. Instead of losing speed, the cyclist kept on accelerating right along with the car, eventually reaching an incredible 226 km/h (140.5 mph).

How could the cyclist go so fast? By taking advantage of the windblocking effect of the car ahead, which cyclists call its *draft*. Behind the large windscreen, it was as if the bicycle were moving through a hole in the air.

TRY THIS

You'll need:
a cardboard toilet-paper roll
a long aluminum foil pan or cookie sheet
matches and permission to use them
two household candles
tape

Creating a draft

1. Light the candles and use some of their melted wax to stick them to the foil pan or cookie sheet about 10 cm (5 inches) apart. Place the toilet-paper roll upright on the pan about 2 cm (an inch) in front of one of the candles and stick it in place with tape.
2. Pull the pan past you with the toilet roll leading. Keep your eyes on the two flames. In what direction are the candle flames blowing?

As the toilet roll moves through the air, it pushes the air aside, creating an area of lower air pressure just behind it. Air from outside that area rushes in to the low-pressure spot, pushing the flame nearest the toilet rolls forward. The flame is drafting.

Imagine that the toilet roll is the leading cyclist in a race and you'll get an idea of why racing cyclists line up one behind the other during a race. They, too, are "drafting."

On level ground overcoming wind resistance consumes about 80 per cent of a bicycle racer's energy. When a racer pulls up close behind another bicycle and drafts, it takes much less energy to push away the air in front of her.

While riding in a draft saves energy, creating the draft uses a lot of energy. That's why the members of cycling teams take turns being the leader.

Cutting through the wind

You'll need:

the same equipment as before, plus

a piece of paper

1. Cut a piece of paper into a rectangle the same height as the toilet roll and long enough to wrap around it and extend about 2 cm (an inch) past on either side. Bend it into a tear-drop shape and tape the ends together. Drop it over the toilet-paper roll with the sharp end pointing toward the candle.

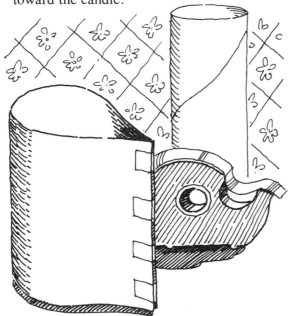

2. Pull the pan past you again. Keep your eyes on the flames. Do they behave any differently this time?

The tear drop shape of the paper allows the air to flow around the toilet roll with less resistance, making it easier for the tube to move through the air. This eliminates the draft behind it. If you were riding behind the leader in a bike race, would you rather follow someone shaped like a toilet roll or a streamlined tear drop?

Stay Out of the Draft

The technique of "drafting" is not something to try on your own. It's difficult to learn and requires proper coaching because it can be dangerous. Such close following often is the reason for those disastrous pile-ups in bike races.

Drafting isn't the only way to decrease wind resistance on a bike. You can also enclose the wheel spokes, as is done on the best racing bicycles. Enclosed spokes (or solid wheels) create less air turbulence, making the bicycle more aerodynamic.

You could also take the whole subject of air resistance lying down. Especially if you're behind the wheel of a "recumbent" bicycle, which resembles a streamlined space pod, since it's enclosed in a shell made of lightweight plastic or fibreglass. You ride these bikes in a reclining position, as if you were steering a sled or toboggan.

WHO'S ON FIRST

How did sports begin? Did a caveman challenge his neighbour one day to a spear-tossing contest? Maybe some cavekids started hitting a rock with a petrified pterodactyl bone?

No one really knows what the first sport was. Archaeologists can only tell us that many sports, such as wrestling, archery, horseback riding, rowing and running, are thousands of years old.

There's no doubt, however, that each sport has a fascinating history. Here are a few of them.

Bowling
You may knock over pins to find out if you're good at bowling. The ancient Germans went bowling to find out if they were good.

In the third and fourth centuries, German men carried a wooden *kegel*, something like the modern bowling pin, which they used for protection, or twirled for exercise. When a man took his kegel to church, he had to set it up at the end of a long hall, then roll a stone ball down the hall to strike the kegel. If he knocked over the kegel, he would be judged clean of sin. If he missed, he obviously needed to devote more time and attention to spiritual matters.

Basketball
One hundred years ago, gymnasiums were dreaded places, home to a dull regimen of calisthenics, Indian club twirling and bar chinning. To fix this situation, Dr. James Naismith, a Y.M.C.A. teacher, invented an indoor game that required players to throw a ball into a hoop high on the wall.

Dr. Naismith never dreamt that basketball would one day be so popular. But he might have had a clue if he'd been around several thousand years ago in Central America. There, *pok-tapok*, an ancient form of basketball, was played in a court surrounded by stone walls and overlooked by sculptures of gods and other religious symbols. The ball was made of rubber and filled with sacred plants, and only the hips, thighs and knees could touch it. The object was to hit the ball into one of the two goals, which were flat stone slabs with a hole cut through the centre.

Weight-Lifting

Weight-lifting is a bold sport, but in ancient Greece it was boulder. That's what the earliest weight-lifters had to pick up. You can still see these huge boulders inscribed with the names of the athletes who lifted them hundreds of years B.C. Picking up heavy rocks spread throughout Europe as a test of manhood, and in some Scottish castles you can still find *clach cuid fir*, or manhood stones.

Modern weight-lifting with barbells and dumbbells became a sensation in the 19th century with the appearance of professional "strongmen" in roving carnivals and vaudeville shows. Strongmen were as celebrated then as movie stars are today, and none more so than the French-Canadian powerhouse, Louis Cyr. Cyr was so strong he could pick up 250 kg (551 pounds) with one finger. In his most famous stunt, Cyr placed his back against the underside of a platform and raised it by exerting the force of his leg and back muscles. On top of the platform stood 18 fat men weighing a total of 1967 kg (4337 pounds)!

Bullfighting

To ancient people, the bull was a powerful symbol of fertility. Farmers used bulls to break up the ground, and when crops sprouted from the tilled soil, the people interpreted this as a sign of the bull's god-like power.

In the centuries that followed, many rituals arose as people tried to gain the bull's mystical force. In Crete, one ritual called for bulls to be drawn into an arena, where young men and women would confront a charging bull, and at the last moment, grab the bull by the horns and somersault onto the animal's back.

In a later rite, a bull was slaughtered over an open pit occupied by bull-worshippers who were drenched in the animal's blood.

The bull-worshipping religions died out long ago, but the rites have survived in the form of a sport.

89

PLAYING WITHOUT SEEING

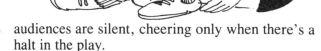

WHAT kind of sport could you play with your eyes closed? How about this one?

You'll need:
a foam ball
some jingle bells
needle and thread or a safety pin or tape or glue
one or more friends
kerchiefs or strips of cloth to use as blindfolds

1. Sew or pin or tape or glue the bells to the foam ball.
2. Close your eyes and tell your friends to close theirs, or put on blindfolds.
3. Roll the ball to a friend and have it rolled back. You need to listen carefully to the bells.
4. If there are several of you, try rolling it back and forth across a circle.

If you find it a challenge to locate and catch a ball with just your ears to guide you, imagine playing a team sport with your eyes closed.

Actually, there is such a sport. It's called goal ball and it's played mainly by people with little or no vision. Like you, they wear blindfolds to give everyone the same chance. There are three players on a side and the teams take turns trying to get a jingling basketball sized ball into one another's goal. Players use all parts of their body to defend the goal, even making spectacular diving saves when necessary.

As you've discovered, it requires concentration and quiet to listen for the jingling, so goal ball audiences are silent, cheering only when there's a halt in the play.

If you think playing ball's difficult when you can't see, how about running a race? You not only have to concentrate on moving fast but you have to do it without banging into anything. One solution is to run holding on to a tether held by a sighted person for guidance. But when you're holding a rope, you can't use both arms for running. And arms are important for balance. Besides, the sighted person then leads.

The other solution is to run in front of a sighted person who guides you with taps on your hip or inside your elbow. That way your body is free to run and you can go at your own pace.

One sport in which a blind athlete must run without physical guidance is the long jump. The only difference in the sport for those who can't see is that the take-off area is a metre (yard) long instead of being marked by a board. It's covered with talcum powder and the jump is measured from the athlete's last footprint in the powder. The athlete starts at the take-off area, then walks back to the start, counting off the paces. The athlete must run alone down the runway and take off without outside cues. Some blind runners use callers, people who stand at the far end of the pit and yell out a repeated sound which the jumper can use for orientation, but they are not allowed to tell the athlete when to jump.

TRY THIS

Take a guided tour
You'll need:
a friend
a blindfold
some small pieces of paper
a clear, grassy area

1. Lay out a course to follow by dropping pieces of paper on the ground. (Be sure to pick them up when you're through.)
2. Put on the blindfold.
3. Walk the course, with your friend behind you, tapping you on your hip or elbow to guide you. How fast can you walk?

Running blind

How hard is it to run in a straight line without looking and without guidance?

You'll need:
a long-jump pit (look for one on the athletic field at your local high school)
two friends
a blindfold

1. Position a friend on either side of the runway to tell you when to stop or catch you if you go off course.
2. Stand at the starting line and put on the blindfold.
3. Run to the take-off point.

How quickly do you go off course? In which direction? Which hand do you usually use? You'll tend to veer in that direction.

So, is the *Guinness Book of World Records* all filled? Do you figure that by the time you're ready to compete, people will have gone as fast, as high, as far as they can? Guess again! Records are broken every day, and not just the kind you put on the turntable. Take swimmer Mark Spitz for instance.

At 18 years of age Mark Spitz won two gold medals, a silver and a bronze at the Mexico City Olympics. In the following four years, he broke 28 world records in the freestyle and butterfly. By the next Olympics, in Munich, Germany, he was without doubt the best swimmer in the world. He won seven gold medals, the most ever won by a single competitor in one Olympiad. Made a mark on history that could never be surpassed, right? Wrong.

Since Spitz retired, swimmers around the world have gotten remarkably faster, smashing records almost by the month. Four Olympics later, Mark Spitz's gold-medal times would not even qualify for Olympic try-outs.

What happened? Did someone slip dolphins into bathing suits? Why are today's swimmers so much swifter? Well, for one thing, they've transformed themselves from canoes to motor boats. The old swimming style required that swimmers pull their arms through the water in a straight line beneath the body, just like canoe paddles. Today's swimmers pull their arms through the water in an S-shaped path that resembles the movement of a motor-boat propeller, pushing them ahead faster.

Swim wear has gotten "faster," too. Swimsuits made of special nylon or lycra fabrics let a swimmer slide through the water more easily. And

improved goggles reduce eye irritation from pool chemicals, allowing swimmers to work out longer.

Swimmers are swifter, too, because the pools where they compete are easier to race in, having been specially built to remove turbulence. If you've ever swum in a public pool you know how choppy the water can get when it's crowded. In a swim meet, this turbulence can really hold you back. But modern pools have lane dividers strung with special discs that keep small waves from forming on the surface. Also, the gutters at the edges of the new pools are contoured so that they swallow choppy water and don't allow it to bounce back into the pool.

Finally, modern pools are much deeper, which reduces underwater turbulence bouncing back from the bottom. The next time you watch a swim meet, notice how smooth the waters are. Try keeping them that still in your bathtub!

Changes in athletic style, clothes, equipment and training methods have not only benefitted swimming, they have also changed other sports. The "four-minute mile" used to be the barrier no runner could break; now, it seems slow. The 50-goal scorer in hockey was once an almost mythical being; now you just about have to score 50 goals before anyone will even collect your bubble-gum card.

Sport is one way that people set up our ideals of what we can do physically. And science keeps pushing back the boundaries. There'll be lots of room for you inside.

ACTIVITIES AND EXPERIMENTS

INDEX